Advances in

Geosciences

Volume 4: Hydrological Science (HS)

ADVANCES IN GEOSCIENCES
A 5-Volume Set
Editor-in-Chief: Wing-Huen Ip *(National Central University, Taiwan)*

Volume 1:   Solid Earth (SE)
             ISBN 981-256-985-5

Volume 2:   Solar Terrestrial (ST)
             ISBN 981-256-984-7

Volume 3:   Planetary Science (PS)
             ISBN 981-256-983-9

Volume 4:   Hydrological Science (HS)
             ISBN 981-256-982-0

Volume 5:   Oceans and Atmospheres (OA)
             ISBN 981-256-981-2

Advances in
Geosciences

Volume 4: Hydrological Science (HS)

Editor-in-Chief

Wing-Huen Ip

National Central University, Taiwan

Volume Editor-in-Chief

Namsik Park

Dong-A University, Korea

World Scientific

NEW JERSEY · LONDON · SINGAPORE · BEIJING · SHANGHAI · HONG KONG · TAIPEI · CHENNAI

Published by

World Scientific Publishing Co. Pte. Ltd.
5 Toh Tuck Link, Singapore 596224
USA office: 27 Warren Street, Suite 401-402, Hackensack, NJ 07601
UK office: 57 Shelton Street, Covent Garden, London WC2H 9HE

British Library Cataloguing-in-Publication Data
A catalogue record for this book is available from the British Library.

ADVANCES IN GEOSCIENCES
A 5-Volume Set
Volume 4: Hydrological Science (HS)

ISBN 981-256-456-X (Set)
ISBN 981-256-982-0 (Vol. 4)

Typeset by Stallion Press
Email: enquiries@stallionpress.com

Printed by FuIsland Offset Printing (S) Pte Ltd, Singapore

EDITORS

Editor-in-Chief: Wing-Huen Ip

Volume 1: Solid Earth (SE)
Editor-in-Chief: Chen Yuntai
Editors: Zhong-Liang Wu

Volume 2: Solar Terrestrial (ST)
Editor-in-Chief: Marc Duldig
Editors: P. K. Manoharan
    Andrew W. Yau
    Q.-G. Zong

Volume 3: Planetary Science (PS)
Editor-in-Chief: Anil Bhardwaj
Editors: Francois Leblanc
    Yasumasa Kasaba
    Paul Hartogh
    Ingrid Mann

Volume 4: Hydrological Science (HS)
Editor-in-Chief: Namsik Park
Editors: Eiichi Nakakita
    Chulsang Yoo
    R. B. Singh

Volume 5: Oceans & Atmospheres (OA)
Editor-in-Chief: Hyo Choi
Editors: Milton S. Speer

REVIEWERS

The Editors of Volume 4 would like to acknowledge the following referees who have helped review the papers published in this volume:

H. Achyutha
Shakeel Ahmed
Dinaud Alkema
Song Baoping
Berndtsson
Zaake Benon
A. Ghosh Bobba
Carl Bruch
Chen
Ramesh Chand
Yuichiro Fujita
Ryo Fujikura
J. S. Gardner
Hoxhaj
Sung Hun Hong
Akira Kawamura
So Kazama
Toshiharu Kojiri
R. H. Kripalani
Sung Jun Kim
Sangdan Kim
Sangjun Kim
S. I. Lee
Shie-Yui Liong
Gwo-Fong Lin
Shashi Mathur
Nandalal
Eiichi Nakakita
Mikiyasu Nakayama
Palanichamy
C. K. Park

Namsik Park
Tilak Priyadarshana
S. N. Rai
A. Rastogi
S. Sen Roy
Jaekyung Rho
A. I. Rugumayo
V. K. Saxena
R. B. Singh
Vijay P. Singh
Dimitri P. Solomatine
Jun, Kyung Soo
M. Sekhar
Sha-Chul Shin
Z. Song
S. H. Song
Sri Srikanthan
P. Sunderambal
Yasuto Tachikawa
Changyuan Tang
Yasuhiro Takeman
Keiichi Toda
Kalanithy Vairavamoorthy
Z. Wang
Wu
Chulsang Yoo
Yong Nam Yoon
Jae-Young Yoon
Junichi Yoshitani
Ziya Zhang

Contents

ON THE LINKAGE OF LARGE-SCALE CLIMATE VARIABILITY WITH LOCAL CHARACTERISTICS OF DAILY PRECIPITATION AND TEMPERATURE EXTREMES: AN EVALUATION OF STATISTICAL DOWNSCALING METHODS

VAN-THANH-VAN NGUYEN[*,§], TAN-DANH NGUYEN[*]

and PHILIPPE GACHON[†]

*Department of Civil Engineering and Applied Mechanics
McGill University, Montreal, Quebec, Canada H3A 2K6
§van.tv.nguyen@mcgill.ca

†Environment Canada and OURANOS Consortium
Montreal, Quebec, Canada

This paper provides an overview of various downscaling methods that could be used for assessing the potential impacts of climate change and variability on hydrological regime. In general, two broad categories of the downscaling techniques currently exist: dynamical downscaling (DD) and statistical downscaling (SD) procedures. While neither DD methods nor SD methods were found to outperform the other, the SD techniques have several practical advantages. Hence, two popular SD methods based on the statistical downscaling model (SDSM) and the stochastic weather generator (LARS-WG) were selected for testing their feasibility in the simulation of two fundamental hydrologic processes: daily precipitation time series and daily temperature extremes. Results of the evaluation using available climate data in the Montreal region (Quebec, Canada) have indicated that both models were able to describe accurately the basic statistical properties of daily maximum and minimum temperatures at local sites. However, none of these models appears to be able to simulate well the statistical properties of the daily precipitation processes.

1. Introduction

General circulation models (GCMs) have been recognized to be able to represent reasonably well the main features of the global atmospheric circulation, but these models so far could not reproduce well details of regional climate conditions at temporal and spatial scales of relevance to hydrological impact studies. Hence, there is a need to develop tools for downscaling GCM predictions of climate change to regional scales. Of particular importance for the management of water resources systems are those tools dealing with the linkage of the large-scale climate variability to the historical observations of the surface parameters of interest (e.g. precipitation and temperature). If this linkage could be established, then the projected change of climate conditions given by a GCM could be used to predict the resulting change of the selected surface parameters. The required linkage could be developed using downscaling methods.

In general, two broad categories of downscaling procedures currently exist: dynamical downscaling (DD) and statistical downscaling (SD).[1] DD procedures are mainly based on regional climate models (RCMs) that describe the climate processes using fundamental conservation laws for mass, energy and momentum. DD methods contain thus more complete physics than SD techniques. However, the more complete physics significantly increases computational cost, which limits the simulation of a climate by RCMs to typically a single realization. On the other hand, SD approaches are relatively fast and much less expensive. These advantages of the SD allow the users to develop a large number of different climate realizations and thus to be able to quantify the confidence interval of simulated climate variables. In addition, SD methods can directly account for the observed weather data available at the study site. The results are hence more consistent with the local climate conditions.

Some recent comparisons of DD and SD techniques for climate impact studies[1-3] have indicated that neither technique was consistently better than the other. In particular, based on the assessment of the climate change impacts on the hydrologic regimes of a number of selected basins in the United States, Gutowski et al.[3] have found that these two methods could reproduce some general features of the basin climatology, but both displayed systematic biases with respect to observations as well. Furthermore, a main finding from this study was that the assessment results were dependent on the specific climatology of the basin under consideration. Hence, it is necessary to test different downscaling methods in order to find the most suitable approach for a particular region of interest. However, it has been widely recognized that SD methods offer several practical advantages over DD procedures, especially in terms of flexible adaptation to specific study purposes, and inexpensive computing resource requirements.[4,5]

In view of the above-mentioned issues, the main objective of the present study is to perform a critical assessment of the adequacy of various existing SD techniques to find the most suitable procedure for hydrological impact studies. Of particular interest is the ability of SD techniques to simulate accurately the characteristics of precipitation and temperature extremes since these two parameters are the main components of the hydrologic cycle.

In this study, the feasibility of two popular downscaling methods, namely the statistical downscaling model (SDSM) and the stochastic weather generator (LARS-WG) model, were assessed using daily precipitation and temperature extreme data available in the southern Quebec region in Canada for the 1961–1990 period. In general, it was found that both models were able to describe accurately the basic statistical properties of daily temperature extremes at local sites. However, none of these models appears to be able to simulate well the statistical properties of daily precipitation processes.

2. Evaluation of Statistical Downscaling Methods

As mentioned above, because of various practical advantages of SD methods over DD procedures, two popular SD techniques based on the SDSM[6] and the LARS-WG[7] model have been selected in this study for testing their feasibility in the simulation of daily precipitation and extreme temperature processes for the Montreal region in Quebec, Canada.

The SDSM is best described as a hybrid of the stochastic weather generator and regression-based methods. The model permits the spatial downscaling of daily predictor-predictand relationships using multiple linear regression techniques and generates "synthetic predictand" that represents the generated local weather. Further details of SDSM are provided by Wilby *et al.*[6] The LARS-WG model[7] produces synthetic daily time series of maximum and minimum temperatures, precipitation, and solar radiation. The model uses input observed daily weather for a given site to determine the parameters of specifying probability distributions for weather variables as well as the correlations between these variables. As mentioned above, LARS-WG utilizes semi-empirical distributions for the lengths of wet and dry day series, daily precipitation and daily solar radiation. The generation procedure to produce synthetic weather data is then based on selecting values from the appropriate distributions using a pseudo-random number generator.

2.1. *Data*

Two data sets are used: station data and NCEP re-analysis data. Station data include observed daily precipitation, daily maximum temperature (t_{max}), and daily minimum temperature (t_{min}) for the period 1961–1990 recorded at four stations in the greater region of Montreal (Quebec,

Table 1.   List of atmospheric variables in NCEP re-analysis data.

Variable	Level of measurement		
Mean sea level pressure	Surface		
Airflow strength	Surface	500 hPa	850 hPa
Zonal velocity	Surface	500 hPa	850 hPa
Meridional velocity	Surface	500 hPa	850 hPa
Vorticity	Surface	500 hPa	850 hPa
Wind direction	Surface	500 hPa	850 hPa
Divergence	Surface	500 hPa	850 hPa
Specific humidity	Near surface	500 hPa	850 hPa
Geopotential height		500 hPa	850 hPa

Canada): Dorval, Drummondville, Maniwaki, and McGill. NCEP re-analysis data are composed of 24 daily atmospheric variables for the same period which are selected for the grid box covering each of the stations considered (see Table 1).

2.2. *Procedures*

SDSM uses NCEP reanalysis data as predictors and station data as predictands, whereas LARS-WG model requires only the station data. Data for the 1961–1975 period were used for the models' calibration step, and those of 1976–1990 for the models' validation. After calibration, the calibrated models are run with the models' parameters and climate conditions for the period 1961–1975 to generate 100 series of local weather data, each series has 15 years of length. At the validation step, the calibrated models are run with the models' parameters and climate conditions for the period 1976–1990 to generate 100 series of local weather data, each of 15 years long. The outputs are statistically analyzed and compared to the statistics of observed data for the same period to evaluate the models' performance. Table 2 presents the evaluation statistics and indices for comparing the performance of the SDSM and LARS-WG models. In addition, to compare the accuracy of the simulation results given by these two models, a scoring technique is used. In this scoring technique, when comparing the bias of an evaluation index, score 1 will be given to the model that has larger bias and score 0 to the one having smaller bias; if the biases of an index of the two models are the same, score 0 will be given to both models. Finally, the model with a larger total score implies that it is less accurate than the one with a smaller total score.

Table 2. Evaluation statistics and indices.

No.	Code	Unit	Time scale	Description
1	Prcp1	%	Season	Percentage of wet days (daily precipitation $\geq 1\,\text{mm}$)
2	SDII	mm/r.day	Season	Sum of daily precipitation/number of wet days
3	CDD	days	Season	Maximum number of consecutive dry days (daily precipitation $< 1\,\text{mm}$)
4	R3days	mm	Season	Maximum 3-day precipitation total
5	Prec90p	mm	Season	90th percentile of daily precipitation amount
6	Precip_mean	mm/day	Month	Sum of daily precipitation in a month/number of days in that month
7	Precip_sd	mm	Month	Standard deviation of daily precipitation in a month
8	DTR	°C	Season	Mean of diurnal temperature range
9	FSL	days	Year	Frost season length: $t_{\text{mean}} < 0°\text{C}$ more than 5 days and $t_{\text{mean}} > 0°\text{C}$ more than 5 days
10	GSL	days	Year	Growing season length: $t_{\text{mean}} > 5°\text{C}$ more than 5 days and $t_{\text{mean}} < 5°\text{C}$ more than 5 days
11	Fr_Th	days	Month	Days with freeze and thaw cycle ($t_{\text{max}} > 0°\text{C}$ and $t_{\text{min}} < 0°\text{C}$)
12	Tmax90p	°C	Season	90th percentile of daily t_{max}
13	Tmax_mean	°C	Month	Mean of daily t_{max} in a month
14	Tmax_sd	°C	Month	Standard deviation of daily t_{max}
15	Tmin10p	°C	Season	10th percentile of daily t_{min}
16	Tmin_mean	°C	Month	Mean of daily t_{min} in a month
17	Tmin_sd	°C	Month	Standard deviation of daily t_{min}
18	Tmoy_mean	°C	Month	Mean of daily t_{mean} in a month
19	Tmoy_sd	°C	Month	Standard deviation of daily t_{mean}

3. Results

3.1. *Model calibration*

Results of the calibration of the SDSM for simulating daily precipitation and temperature extremes showed that the daily precipitation predictand requires significant predictors such as zonal velocities, meridional velocities, specific humidities, geopotential height, and vorticity, while both daily maximum (t_{max}) and minimum (t_{min}) temperatures predictands require geopotential heights and specific humidities at all levels. The coefficients of determination (R^2) after calibration of the SDSM for t_{max} and t_{min} were very high, from 0.714 to 0.785, respectively, while the value for precipitation was very low, ranging from 0.062 to 0.098. This would indicate the difficulty in finding significant climate variables from the NCEP data that could explain well the variability of daily precipitation. Similar difficulty was found for the LARS-WG in the simulation of daily precipitation characteristics even though the bias values of different indices for LARS-WG are generally smaller than those given by the SDSM. Tables 3 and 4 show the total bias scores produced by the SDSM and LARS-WG models for daily precipitation and for daily temperature extremes, respectively. It was found that the LARS-WG model was able to produce daily precipitation statistics in closer agreement with those of the observed data, while SDSM

Table 3.   Total bias scores for daily precipitation simulation for 1961–1975 calibration period.

Evaluation indices	SDSM	LARS-WG
SDII	13	3
Prec90p	10	6
Prcp1	15	1
CDD	10	1
R3days	12	3
Prec-moy	33	11
Prec-std	38	9
Total	131	34

Table 4.   Total bias scores for extreme daily temperature simulation for 1961–1975 calibration period.

Evaluation indices	SDSM	LARS-WG
Tmax-moy	2	46
Tmax-std	10	38
Tmax90p	11	5
Tmin10p	5	10
Tmin-moy	2	45
Tmin-std	29	19
Tmean-moy	1	47
Tmean-std	21	27
DTR	7	3
Fr-Th	11	11
FSLs	3	1
GSL	2	2
Total	104	254

provided daily extreme temperature statistics that were more accurate than the LARS-WG.

3.2. *Model validation*

Results of the SDSM and LARS-WG for the 1976–1990 validation period at the four selected sites indicated that neither of these models was able to generate accurately the statistics of daily precipitation. For instance, bias for the mean of daily precipitation amount (SDII) ranges from −0.39 to +1.94 mm for SDSM, and −0.32 to +0.44 mm for LARS-WG; and for the 90th percentile of daily precipitation (*Prec90p*) from −2.4 to +3.12 mm for SDSM, and from −1.93 to +3.37 mm for LARS-WG. Table 5 shows the total bias scores of the two models in the simulation of daily precipitation process. It can be seen that LARS-WG can reproduce more accurate observed statistics of daily precipitation than the SDSM.

Regarding the simulation of daily temperature extremes, both SDSM and LARS-WG could provide an accurate description of many daily maximum and minimum temperature characteristics. For instance, the bias values for the extreme statistic *Tmax90p* are from −0.9°C to +0.82°C for SDSM, and from −1.55°C to +1.03°C for LARS-WG. Similarly, for the extreme index *Tmin10p* the bias values are from −0.18°C to +0.98°C for SDSM and from −0.46°C to +0.48°C for LARS-WG. The small bias values indicate the accuracy of both models in the simulation of the extremes of daily maximum and minimum temperatures. Similar results were found for other evaluation statistics. Furthermore, Table 6 shows the total bias scores for SDSM and LARS-WG. It can be seen that both models could provide a comparable performance in the simulation of extreme daily temperature processes.

Table 5. Total bias scores for daily precipitation simulation for 1976–1990 validation period.

Evaluation indices	SDSM	LARS-WG
SDII	14	1
Prec90p	9	6
Prcp1	14	2
CDD	6	4
R3days	14	2
Prec-moy	40	8
Prec-std	38	9
Total	135	32

Table 6.   Total bias scores for extreme daily temperature simulation for 1976–1990 validation period.

Evaluation indices	SDSM	LARS-WG
Tmax-moy	16	32
Tmax-std	31	13
Tmax90p	9	7
Tmin10p	9	7
Tmin-moy	22	25
Tmin-std	24	22
Tmean-moy	14	33
Tmean-std	33	12
DTR	8	7
Fr-Th	8	10
FSLs	0	3
GSL	1	3
Total	175	174

4. Summary and Conclusions

In this study, because of various practical advantages of SD methods over DD procedures, two popular SD techniques based on the LARS-WG and the SDSM models have been selected for testing their ability to simulate daily precipitation and extreme temperature series for four raingage stations in the Montreal region in Quebec, Canada. Calibration of the SDSM suggested that local precipitation was mainly related to the large-scale climate variables such as the zonal velocities, meridional velocities, specific humidities, geopotential height, and vorticity, while the local maximum and minimum temperatures were strongly related to the geopotential heights and specific humidities at all levels.

The comparison between some selected statistics of observed weather data and those of weather data generated by the two models indicates that the LARS-WG model can provide the daily precipitation statistics more comparable to those of the observed data than the SDSM. However, both models were unable to reproduce accurately these observed statistics. The SDSM and LARS-WG were found to be able to describe adequately the observed statistics of daily temperature extremes, and the SDSM was found to be somewhat more accurate than the LARS-WG. In terms of practical applications, calibration of the LARS-WG model is much simpler than that of the SDSM. The calibration of the SDSM is based on a complex procedure in order to be able to establish successfully the good relationships between large-scale predictor variables and the surface weather variables at a local site.

Further studies are planned to compare the performance of these two models using data from other sites with different climatic conditions. In addition, the performance of these two models should be evaluated using data from GCMs in order to be able to assess the reliability of generated future climate scenarios for a local site.

References

1. B. Yarnal, A. C. Comrie, B. Frakes and D. P. Brown, Developments and prospects in synoptic climatology. *International Journal of Climatology* **21** (2001) 1923–1950.
2. LO. Mearns, I. Bogárdi, I. Matyasovszky and M. Palecki, Comparison of climate change scenarios generated from regional climate model experiments and statistical downscaling, *Journal of Geophysical Research* **104** (1999) 6603–6621.
3. W. J. Gutowski, Jr., R. L. Wilby, L. E. Hay, C. J. Anderson, R. W. Arritt, M. P. Clark, G. H. Leavesley, Z. Pan, R. Da Silva and E. S. Takle, Statistical and dynamical downscaling of global model output for US national assessment hydrological analyses, *Proceedings of the 11th Symposium on Global Change Studies*, Long Beach, CA, January 9–14, 2000.
4. C.-Y. Xu, From GCMs to river flow: A review of downscaling methods and hydrologic modeling approaches. *Progress in Physical Geography* **23**, 2 (1999) 229–249.
5. C. Prudhomme, N. Reynard and S. Crooks, Downscaling of global climate models for flood frequency analysis: Where are we now? *Hydrological Processes* **16** (2002) 1137–1150.
6. R. L. Wilby, C. W. Dawson and E. M. Barrow, SDSM – A decision support tool for the assessment of regional climate change impacts. *Environmental Modelling and Software* **17** (2002) 147–159.
7. M. A. Semenov and E. M. Barrow, Use of stochastic weather generator in the development of climate change scenarios, *Climatic Change* **35** (1997) 397–414.

ENVIRONMENTAL ISOTOPES OF PRECIPITATION, GROUNDWATER AND SURFACE WATER IN YANSHAN MOUNTAIN, CHINA

K. AJI[*,¶,‖], T. CHANGYUAN[*], A. KONDOH[†], S. XIANFANG[‡] and Y. SAKURA[§]

[*]*Graduate School of Science and Technology, Chiba University, Japan*
[†]*Center for Environmental Remote Sensing, Chiba University, Japan*
[‡]*Institute of Geographical Sciences & Natural Resources Research,*
Chinese Academy of Science (CAS), Beijing, China
[§]*Department of Earth Sciences, Faculty of Science, Chiba University, Japan*
[¶]*kaisar@graduate.chiba-u.jp*
[‖]*kaisar527@yahoo.com*

Analysis of stable isotopes in precipitation, groundwater and surface water in Yanshan Mountain of the North China Plain (NCP) was carried out to identify the hydrological processes. In Yanshan Mountain the δD and δ^{18}O values of water show a very distinct pattern, with wide ranged isotope values from high altitude to mountain hill. The distribution of δD and δ^{18}O in precipitation, groundwater and surface water suggested that the groundwater was recharged by the direct diffuse recharge of precipitation, as well as from the surface water. Meanwhile the ground water recharged to the surface water at different region with different form.

1. Introduction

Depended heavily upon groundwater resources, the NCP is the China's leading industrial and agricultural region that produces more than 50 percent of the nation's wheat and 33% of its maize, and is the home to more than 200 million people (State Statistics Bureau 1999[6]). In recent decades the aquifers in the NCP have been rapidly developed for urban and industrial water supply, as well as agricultural irrigation. In fact, groundwater in the NCP is used at a rate much higher than that with which the aquifers are filled.

Previous hydrochemical investigation of the NCP were focused on the pleoclimatic change.[1] Several studies have shown that the groundwater in the area near Taihang Mountain, which was taken from comparatively shallow wells, is quite young and belongs to the recharge area for the NCP. The purpose of this paper is to explain the hydrological characteristics in the Yanshan Mountain, the north part of the NCP, where study has previously been limited.

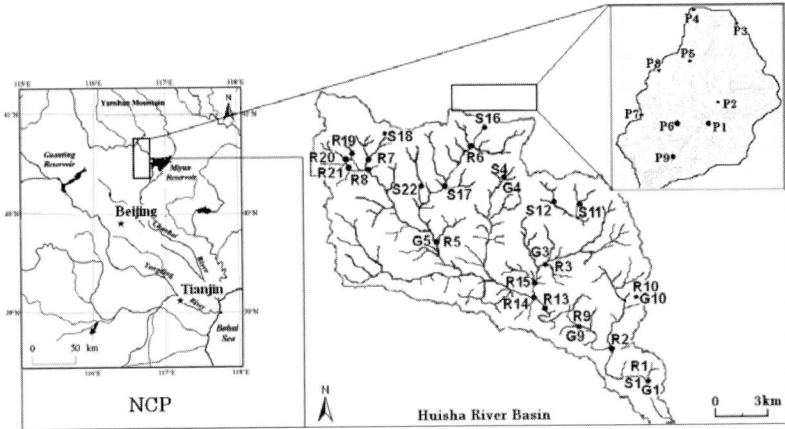

Fig. 1.   Sampling location in the study area of the North China Plain.

2. Hydrogeological Settings and Methods

Located at lower part of Yanshan Mountain in the NCP, the Huaisha River Basin (Fig. 1) is typical basin in a hilly area as large as $155 \, km^2$. It belongs to the littoral and semi-arid climatic zone, and annual rainfall ranges from 400 to 600 mm. Mean potential evaporation ranges from 1,100 to 1,800 mm.[2]

The east–west trending Yanshan Mountain is underlies by Archian crystalline basement rocks.[3] The bedrock is composed of Archaeozoic gneiss and Proterozoic carbonate. Generally, groundwater is recharged in Taihang and Yanshan Mountains then moves eastward and southward to the Bohai Sea.[1] Most recharge is from infiltration of precipitation near the outcrop area around the mountains.

Precipitation, groundwater (shallow) and surface water (river and spring water) were collected in September 2003. In several point in the study area collected both groundwater and surface water. The δD and $\delta^{18}O$ of samples were measured by the mass spectrometer (Delta S Thermoqest) in the laboratory of Chiba University.

3. Results and Discussions

3.1. *Stable isotopes of precipitation, groundwater, and surface water*

Isotope and precipitation data are presented in Table 1. Isotope composition of precipitation ranged from $-8.38‰$ to $-10.51‰$ for $\delta^{18}O$ and -70.3

Table 1.   Isotopic characteristics of water samples in the study area.

Point	Type	$\delta^{18}O$ (‰)	δD (‰)	d-excess (‰)	Point	Type	$\delta^{18}O$ (‰)	δD (‰)	d-excess (‰)
R1	R	−7.84	−65.9	−3.18	R13	R	−9.02	−69	3.16
G1	GW	−8.46	−69.8	−2.12	R14	R	−8.78	−64.6	5.64
S1	S	−7.67	−65.4	−4.04	R15	R	−8.65	−68.2	1
R2	R	−7.79	−64.3	−3.28	S16	S	−9.8	−70.1	8.3
R3	R	−8.47	−69.6	−1.84	S17	S	−10.1	−69.6	11.2
G3	GW	−8.96	−70.1	1.58	S18	S	−9.22	−66.8	6.96
S4	S	−8.64	−71.2	−2.08	R19	R	−8.06	−62	2.48
G4	GW	−8.62	−72.7	−3.74	R20	R	−8.3	−62.2	4.2
R5	R	−9.15	−74.1	−0.9	R21	R	−8.28	−64.5	1.74
G5	GW	−8.48	−72.1	−4.26	S22	S	−9.34	−68.9	5.82
R6	R	−8.72	−71.5	−1.74	P1	Rain	−8.79	−72.5	−2.18
R7	R	−9.1	−68.2	4.6	P2	Rain	−8.84	−75.5	−4.78
R8	R	−8.4	−62.1	5.1	P3	Rain	−9.57	−77.1	−0.54
R9	R	−8.21	−62.7	2.98	P4	Rain	−10.51	−82.4	1.68
G9	GW	−8.6	−61.4	7.4	P5	Rain	−9.4	−72.5	2.7
R10	R	−7.77	−62	0.16	P6	Rain	−8.72	−74.4	−4.64
G10	GW	−8.67	−62	7.36	P7	Rain	−8.63	−71.8	−2.76
S11	S	−8.98	−63.7	8.14	P8	Rain	−9.01	−75.2	−3.12
S12	S	−8.95	−71.7	−0.1	P9	Rain	−8.38	−70.3	−3.26

Note: GW: groundwater, R: river water, S: Spring water.

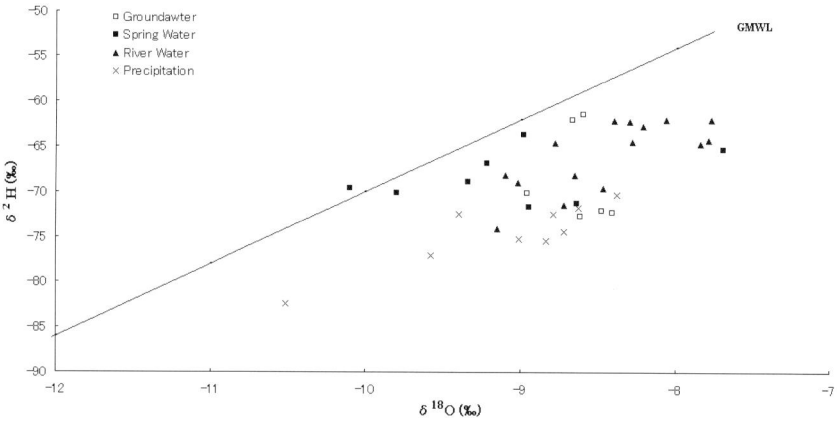

Fig. 2.   Relations between δD and $\delta^{18}O$ in water samples of the Huaisha River Basin.

to −82.4‰ for δD values in the study area. The precipitation samples plotted to the right of the global meteoric water line (GMWL) of Craig[4] with the
isotopic composition of precipitation undergoing evaporation tends to fall
on lines of lower slopes (Fig. 2). The enriched isotopic values documented

in the lower slope of the study area could be result of the evaporation. The mean deuterium excess d $(= \delta D - 8\delta^{18}O)$ is $-1.88‰$ of precipitation in the study area. Precipitation affected by evaporation and tends to have d values (d-excess) lower than 10.[5] The most depleted value occurred at highest altitude in point P4 which $-10.51‰$ for $\delta^{18}O$ and $-82.4‰$ for δD, and enriched value at lowest altitude in point P9 which $-8.38‰$ for $\delta^{18}O$ and $-70.3‰$ for δD due to the altitude effect.

The isotopic composition of the groundwater ranged from $-8.96‰$ to $-8.46‰$ for $\delta^{18}O$ and -72.7 to $-61.4‰$ for δD in the study area. All groundwater samples plotted to the right of the global meteoric water line (Fig. 2). This is also characteristic future of mountain area in the NCP, which is probably due to isotopic enrichment caused by evaporation that occurred before and during infiltration in the recharge area, and similar observation were confirmed by Chen et al.[1] in Taihang Mountain. In the upper part, the groundwater sample was enriched at point G5. The groundwater reached at the middle part of the study area the values was depleted. In this part groundwater mainly recharged by spring water, point S11, S12, and G3 observed very similar values which -8.98, -8.95, and $-8.96‰$ for $\delta^{18}O$, respectively. The more enriched value was found in point G1 ($-8.46‰$ for $\delta^{18}O$) in the lower part and similar to point R9 and P11. It indicates that the groundwater may originate from river water and precipitation with a more enriched low latitude area and should reflect the isotopic composition of groundwater. The groundwater of the study area was recharged by the direct diffuse recharge of precipitation, as well as recharge from the surface water.

The isotopic composition of the surface water (river and spring water) ranged from $-10.1‰$ to $-7.67‰$ for $\delta^{18}O$ and $-74.1‰$ to $-64.3‰$ for δD in the Huaisha River Basin. The more depleted isotope values occurred in the spring water of point S16 ($-9.8‰$ for $\delta^{18}O$) and point S17 ($-10.1‰$ for $\delta^{18}O$) of upper part, and close to the most depleted precipitation sample at point P4. Probably it was recharged by the precipitation falling in much higher altitude.

3.2. Interaction between groundwater and surface water

In the upper part of the study area, the stable isotope values of river waters such as point R8, R19, R20, and R21 were almost same and similar to the groundwater sample at point G5, which displays enriched value ($-8.48‰$ for $\delta^{18}O$). When the enriched river water reaches an artificial concrete dam

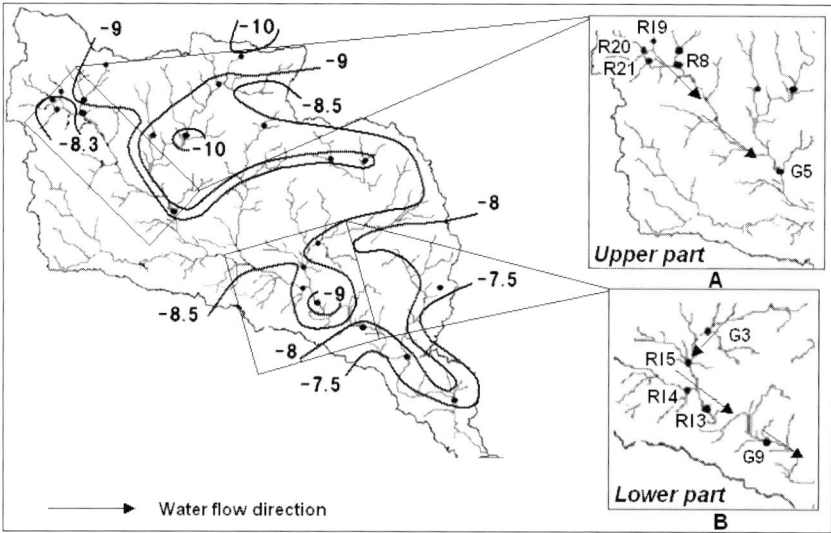

Fig. 3. δ^{18}O distribution in river water of the Huaisha River Basin.

built near the point G5 along the river direction, the enriched river water disappears into underground and recharged the groundwater. The infiltration rates also high in the riverbed, because of the lower runoff the river. This enrichment and similarity at the high altitude area suggested that the groundwater recharged by the river water in this area (Fig. 3A).

In the lower part, the stable isotope values of river water revisal than upper part. The isotope values of river waters at point R15, R14, and R13 were same and depleted than the upper part. This depleted values of river waters similar to groundwater samples at point G3 and point G9 in low-altitude area. This suggested that river water fed by the groundwater in this region (Fig. 3B). Comparing with groundwater at point G1 (δ^{18}O around −8.46‰), the more enriched isotope values were observed in river water at point R1 (−7.84‰ for δ^{18}O) and spring water at point S1 (−7.67‰ for δ^{18}O). Those river water and spring water were collected at the same point of the lower part in the study area with low *d-excess*. The *d-excess* of surface water sample lower than the groundwater sample which result of the evaporation in the lower part. This indicated that groundwater recharged to the surface water, after that the evaporation contributed isotopic enrichment to the surface water in the lower part of the Huaisha River Basin.

4. Conclusions

The range of isotope contents in precipitation, groundwater, and surface water had very good correlation at place to place. The factors such as topography, altitude and seasonality play major role to the spatial distribution of isotope components in mountain region. Interaction between groundwater and surface water is typical characteristics in mountain region and the isotopes is the powerful tool to identify them.

References

1. Z. Chen, J. Qi, J. Xu, H. Ye and Y. Nan, Pleoclimatic interpretation of the past 30 ka from isotopic studies of the deep confined aquifer of the North China Plain, *Applied Geochemistry* **18** (2003) 997–1009.
2. C. F. Liu and P. Y. Wang, The environment significance of isotope composition in groundwater of Hebei Plain, in: *Proceedings of International Workshop on Groundwater and Environment*, Beijing, 16–18 August (1992) (in Chinese).
3. A. Chen, Geometric and kinematic evolution of basement — cored structures: Intraplate orogenesis within the Yanshan Oregen, northern China, *Tectonophysics* **29** (1998) 17–42.
4. H. Craig, Isotopic variations in meteoric waters, *Science* **133** (1961) 1702–1703.
5. Y. Yutsever, Worldwide survey of stable isotopes in precipitation, *Rep. Sec. Isotope hydrol.*, IAEA, 1975.
6. E. Kendy, D. J. Molden, T. S. Steenhuis, C. M. Liu and J. Wang, Policies drain the North China Plain; Agricultural policy and groundwater depletion in Luancheng Country, 1949–2000, Research Report 71, Colombo, Sri Lanka, International Water Management (2003).

SIMULATION KOREAN SUMMER MONSOON RAINFALL WITH NCAR REGIONAL CLIMATE MODEL

G. P. SINGH*,†, JAI-HO OH†,§ and JIN-YOUNG KIM†

*Department of Geophysics, Banaras Hindu University, Varanasi, India
†Department of Environment and Atmospheric Sciences
Pukyong National University, Busan, South Korea
§jhoh@pknu.ac.kr

Some of the important characteristics of Korean summer monsoon circulations and rainfall are examined using the NCAR Regional Climate Model (RegCM3). RegCM3 has been simulated at 27 km horizontal resolution over the east Asia domain for the period from January to December in 2002. The important features of wind and temperature at lower and upper levels over east Asia and precipitation simulated by the model over South Korea are examined in details for different convective parameterization schemes namely, mass flux, a Kuo-type and Emanuel schemes. The monsoon circulation features simulated with RegCM3 are compared with those of the NCEP/NCAR reanalysis and the simulated Korean summer monsoon rainfall (KMR) is validated against the observations from Korea Meteorological Administration (KMA). Validation of simulated precipitation with KMA shows that the use of the Emanuel scheme is more close to the KMA. Characteristic features of the intra-seasonal quasi-biweekly (10–20 days period) and the Madden-Julian (30–60 days period) oscillations have been investigated. Results of intra-seasonal oscillation also indicate that the use of Emanuel convective scheme yields results close to the KMA. Results from wavelet analysis show that the high precipitation in 2002 may be associated with prominent 30–60 days oscillation during the month of August.

1. Introduction

Some of the studies, mainly by Chinese scientists,[1] have mentioned that many differences exist between the monsoon circulations over south and east Asia. This fact suggests that the main component of the east Asian monsoon is likely to be different from the south Asian monsoon systems. Therefore, Asian monsoon systems can be divided into two subsystems namely, south and east Asian monsoons, which are independent to each other. The east Asian summer monsoon system exhibit a large spatial and temporal variability. Best example is a recent case of excess summer monsoon rainfall (KMR) of 2002 over South Korea. A prolonged period of rain and showers occurred from 5 to 18 August 2002 over South Korea and flooding and severe damage in southeastern Korea for more than 10 days.[2] This naturally led to select a case of monsoon of 2002 for our studies.

It is well known that the precipitation over Asia varies considerably from day to day basis. Generally, rain occurs in spell over major parts under the influence of favorable circulation conditions. This periodical rainfall is related to a hierarchy of quasi-periods, namely 3–7 days, 10–20 days, and 30–60 days oscillations. The 3–7 days periodicity is associated with the oscillation of low pressure zone (monsoon trough over India, Changma over Korea, Baiu over Japan, and Meiyu over China), 10–20 and 30–60 days periodicities are associated with the west ward moving waves formed over Bay of Bengal and the globally eastward moving wave numbers 1 and 2 over the tropics,[3] respectively. Most of the observational and modeling studies on the intra-seasonal low-frequency oscillations have focused mainly for south Asian regions, some studies have done for east Asian regions, but a very few studies available over South Korea using regional climate model. Hence, in this study, we have investigated the temporal characteristics of the 10–20 and 30–60 days oscillations to summer precipitation over South Korea using RegCM3 model with different cumulus parameterization schemes. The monsoon of 2002 over South Korea has been selected for this study. We have focused mainly precipitation fields simulated with RegCM3 because of its sensitivity to climate. Section 2 describes the details of model and simulated results are given in Sec. 3. Wavelet analysis of precipitation is presented in Sec. 4 and conclusions are given in Sec. 5.

2. Model Descriptions

The model used in this study is the recent version of NCAR RegCM3.[4–6] The model dynamical core is essentially the same as that of the hydrostatic version of the meso-scale model MM5.[7] The RegCM3 includes the large scale precipitation scheme,[8] the radiative transfer scheme of the global model CCM3[9] and the ocean-atmosphere flux scheme.[10] In this study, we have used three convective parameterization schemes. The first is the simplified Kuo-type parameterization.[7] The second scheme we have used is also extensively used within both the MM5 and RegCM3 modeling frame work. This is a mass flux scheme that includes the moistening and heating effects of penetrative updrafts and corresponding downdrafts. The schemes have two closer assumptions, the so-called Arakawa–Schubert (AS) and the Fritsch–Chappell (FC) type closures. The third scheme is Emanuel[11] (EMU).

The period of RegCM3 simulations are from January to December, 2002 for the purpose of present study. The computation domain covers area approximately 24 N–48 N and 109 E–148 E with a grid point spacing at 27 km. For validation of simulated precipitation, daily rainfall data for 59 stations spread over South Korea, are used during summer monsoon (June–August) in 2002 from Korea Meteorological Administration (KMA). The observed (KMA) precipitation over 59 stations is interpolated using an objective analysis[12] to get better spatial details. For examining the intensity of intra-seasonal oscillations and its temporal variation, Butterworth band pass filter,[13] with half power points at 10 and 20 days for the quasi-biweekly oscillations and at 30 and 60 days for the Madden Julian oscillation (MJO) is applied to the precipitation time series.

3. Simulation of Mean Monsoon Features

The seasonal wind fields simulated by model at lower (850 hPa) and upper (200 hPa) levels have the maximum strength of tropical southerly by 6 m/s and upper level westerly of 25 m/s, respectively. This result show that mean monsoon wind fields at 850 and 200 hPa are in line with the NCEP/NCAR reanalysis. While spatial distribution of seasonal temperature fields (850 hPa) shows that RegCM3 seems to be characterized by cold biases of few degree over some regions. Cold bias in RegCM3 simulation may be caused by the physical parameterizations such as non-local boundary layer, convection and land surface scheme.[14] A comparison of simulated temperature and wind patterns with NCEP/NCAR reanalysis fields indicate that the EMU scheme exhibits a better simulation compared to other convective schemes over east Asia.

Figure 1 shows the precipitation only for the interior of the domain (mainly over South Korea) in order to more clearly illustrate the fine scale topographical induced details. Figure 1(a)–(d) shows the precipitation rate (mm/day) simulated by RegCM3 during summer monsoon season (average of June–August) using various cumulus convective schemes such as AS, FC, a Kuo-type, and EMU in RegCM3. Figure 1(e) shows the observed precipitation rate (mm/day) from KMA. RegCM3 simulation with AS scheme shows the maximum precipitation of 12 mm/day over southern Korea and northeast coast of Korea, while FC scheme shows that the maximum precipitation of 14 mm/day over southwest parts and 12 mm/day over central

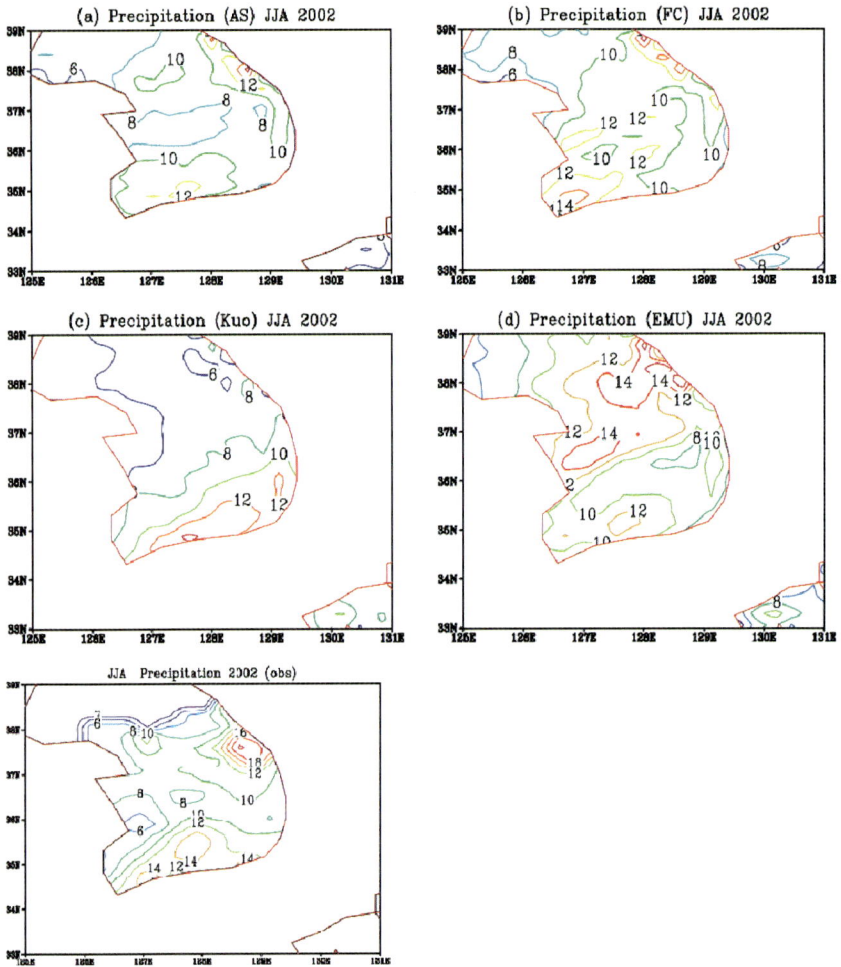

Fig. 1. RegCM3 simulated precipitation using (a) Arakawa–Schubert (AS), (b) Fritsch–Chappell (FC), (c) Kuo, (d) Emanuel (EMU) convective schemes, and (e) observed precipitation from KMA in 2002 during JJA.

parts of Korea. When using Kuo-type scheme, RegCM3 shows the maximum precipitation of 12 mm/day cover good regions of southern parts of Korea. For EMU scheme, the model simulated the maximum precipitation of 14 mm/day over northeast, central west and 12 mm/day over southern parts of Korea. The observed precipitation from KMA shows maximum precipitation of 14 mm/day over southern coast and 16 mm/day over northeast parts of Korea.

Over all, a comparison of observed and RegCM3 simulated precipitation shows that the model captured well the spatial distribution of seasonal precipitation and belts of maximum precipitation. Validation against KMA data sets show that the precipitation simulated with the EMU convection scheme is more realistic compared with other schemes. The RegCM3 simulation locates the maximum precipitation belt over south and east coast and minimum over northwest coast of South Korea, is good agreement with observation (KMA).

4. Wavelet Analysis

Wavelet analysis or wavelet transform (WT) is a powerful mathematical tool well suited for the study of multi-scale non-stationary processes occurring over finite spatial and temporal domains. A detail computational procedure for WT is well described in the website http://ion.researcgsystem.com/IONScript.

The daily rainfall fluctuation and wavelet spectrum are computed from daily precipitation during 2002. Wavelet spectrum for KMA shows the maximum variance after first week of August. It also illustrates that the period of oscillation increases after first week of August, with maximum variance centered around days 65 in 60–80 days time band in KMR (not shown). While model simulated maximum variance is also centered around 65 days in 60–80 days. These maximum values are well reflected in daily precipitation series between 60 and 80 days band for KMA and model (not shown). This analysis shows that EMU convective scheme in RegCM3 are more close to the KMA compared to AS convective schemes.

For examining the intensity of 10–20 and 30–60 days oscillations, Butterworth band pass filter is applied[9] to the precipitation series. The variances in 10–20 and 30–60 days band for KMA are compared with model simulated variance. The model simulation shows that the variance in 10–20 days is approximately twice than those in 30–60 days, agrees well with KMA. The variance obtained to 10–20 and 30–60 days oscillations are 8.7 and 3.0 for KMA, 9 and 4 for model with EMU convective scheme and 13 and 6 for AS scheme in RegCM3, respectively.

Similar patterns of variance for 10–20 and 30–60 days are also obtained[3] on the basis of 23 years (1978–2000) observed data for 20 uniformly distributed stations from KMA. A comparison of model simulated variance in 10–20 and 30–60 days with observed (KMA) shows that the variance simulated with EMU convective schemes in RegCM3 is more close than others schemes.

5. Conclusions

In this paper, RegCM3 model has been integrated over east Asia regions using different cumulus parameterizations schemes. Results indicate that RegCM3 successfully simulates some of important characteristics of the east Asian monsoon circulations, such as tropical southerlies at lower level and westerlies at 200 hPa. Also, the seasonal mean summer monsoon precipitation simulated by RegCM3 is close to the corresponding observed value of KMA when the Emanuel convective scheme is used. The analysis of intra-seasonal oscillation shows that the monsoon precipitation over South Korea as whole is likely to be stronger when 30–60 days oscillations are prominent. In general, Emanuel scheme performed better than AS, FC, and Kuo schemes in simulating both the monsoon circulations and precipitation.

Acknowledgments

This study was funded by the Korea Meteorological Administration Research and Development program under Grant Cater 2006-1101. The authors would like to acknowledge the support from KISTI (Korea Institute of Science and Technology Information) under the 'Sixth Strategic Supercomputing Support Program' with Dr. Lee Sang Min and Dr. Cho Kum Won as the technical supporter. The use of the computing system of the Supercomputing Center is also greatly appreciated. First author wish to acknowledge the KOFST for supporting the visit.

References

1. S. Tao and L. Chen, *Monsoon Meteorology*, eds. C. P. Chang and T. N. Krishnamurti (Oxford Univ. Press, Oxford, 1987), pp. 60–92.
2. Y.-S. Chung, M.-B. Yoon and H.-S. Kim, *Climatic Change* **66** (2004) 151.
3. C.-S. Ryu and R. H. Kripalani, *Korean J. Atmos. Sci.* **5** (2002) 85.
4. F. Giorgi, *Mon. Weather, Rev.* **119** (1991) 2870.
5. F. Giorgi, M. R. Marinucci and G. T. Bates, *Mon. Weather Rev.* **121** (1993a) 2794.
6. F. Giorgi, M. R. Marinucci, G. T. Bates and G. DeCanio, *Mon. Weather Rev.* **121** (1993b) 2814.
7. G. A. Grell, J. Dudhia and D. R. Stauffer, *NCAR Technical Notes*, *NCAR/TN-398+STR*, 21, 1994.
8. J. Pal, E. Smit and E. Eltahir, *J. Geophys. Res.* **29** (2000) 29579.

9. F. Giorgi and L. O. Mearns, *J. Geophys. Res.* **104** (1999) 6335.
10. X. Zeng, M. Zao and R. E. Dickinson, *J. Climate* **11** (1998) 2628.
11. K. A. Emanuel, *J.A.S.* **21** (1991) 2313.
12. S. L. Barnes, *J. Appl. Meteor.* **3** (1964) 396.
13. M. Murakami, *Mon. Weather Rev.* **107** (1979) 1011.
14. F. Giorgi, X. Bi and J. S. Pal, *Climate Dynamics* **22** (2004) 733.

COMBINATION OF CLUSTER ANALYSIS
AND DISCRIMINATION ANALYSIS USING
SELF-ORGANIZING MAP

GWO-FONG LIN*, CHUN-MING WANG

Department of Civil Engineering, National Taiwan University
Taipei 10617, Taiwan
* gflin@ntu.edu.tw

Regionalization is an important technique that uses existing information to extrapolate where the information is required but cannot be obtained. The cluster analysis and discrimination analysis are the two important procedures of regionalization. In this paper, a simple method based on the self-organizing map is proposed to combine the cluster analysis and the discrimination analysis. The advantages of the proposed method are that it can determine the proper number of clusters, reveal the relative relationship of the input patterns and allocate the unknown patterns into known clusters. The design hyetographs of northern Taiwan are analyzed using the proposed method. The clustering results of the design hyetographs of northern Taiwan using the proposed method exhibit homogeneities within clusters and heterogeneities among clusters. Regarding the capability of determining the proper number of clusters, the proposed method is superior to conventional clustering method. The discrimination results also show that the assignments of unknown patterns to known clusters are reasonable using the proposed method. Thus the proposed method can be applied to regionalization to reduce the complexity and the difficulty.

1. Introduction

There is a problem that some specific hydrological informations are required at a certain location where the necessary information cannot be easily obtained often encountered by hydrological engineers. These problems can be solved by using the regionalization. Regionalization is used to extrapolate some hydrological informations from sites at which the hydrological information can be derived to others at which the hydrological informations are required but unavailable.[1] The processes of regionalization may often combine several procedures, including the cluster analysis and the discrimination analysis.[2,3] The cluster analysis is to explore the relative relationships and the grouping of the hydrological factors. However, different methods of cluster analysis applied to the same set of data often

25

lead to different clustering results.[2] The discrimination analysis is to build a model to assign an ungauged site to a known cluster, so that a proper extrapolation model can be selected to estimate the specific hydrological information. Some studies use only cluster analysis for regionalization.[4,5] It is insufficient to allocate an ungauged site to a known cluster. Therefore, the discrimination analysis is necessary for regionalization.

Artificial neural network is now a popular tool to deal with massive and complex data to derive useful information. The artificial neural network used herein is self-organizing map (SOM) proposed by Kohonen.[6] SOM is a competitive and unsupervised network. Mangiameli et al.[7] compared SOM with other seven hierarchical clustering methods. The result of the study shows that the performance of SOM in clustering messy data is better than it of the other seven hierarchical clustering methods. Michaelides et al.[8] adopted the SOM to classify the rainfall variability to provide prototype classes of weather variability. The result shows that SOM can detect much more detail of rainfall variability than hierarchical clustering methods.

The purpose of this paper is to propose a simple procedure that can classify the hydrological factors and simultaneously allocate an ungaued site to a known cluster properly.

2. Method

SOM is known as one kind of artificial neural networks. The essential mechanism of SOM is the competitive and unsupervised learning process in which the neurons of the network compete each other to be activated. The attractive capabilities of SOM are to map the high-dimensional input patterns into a lower-dimensional output space and to preserve the topological relations of input patterns. Readers can refer to Kohonen's[6] and Haykin's[9] books for more details.

A fascinating feature of SOM is that the relationships of input patterns can be stored within the network, after the training process is completed. Since the relationships of input patterns can be stored, an SOM-based cluster and discrimination analysis (SOMCD) method is proposed.

After the SOM training is done, feeding the SOM with all input patterns that have learned by the SOM can lead to the feature map. The way to obtain the feature map is to label all winning neurons in the output space with the identities of corresponding input patterns.

If a neuron responds to a specific input pattern, the neuron is called the image of the specific input pattern or the neuron is "imaged" by the specific input pattern. The density map can be obtained by applying the following equation to a trained SOM:

$$N = \text{Num}(i_j), \quad j = 1, 2, \ldots, l, \tag{1}$$

where N is an integer, i is the neuron of the trained SOM, and Num( ) is a function counting the number of the neuron i "imaged" by certain input patterns. Every pattern in the input space has only one image, but one neuron can be the image of many input patterns. The density map can reveal the grouping of input patterns. From Eq. (1), the density map can be obtained easily by labeling each grid of the map with the integer N. Suppose that the number of each grid is the "elevation" of the density map. Then the grouping of input patterns is shown by certain isolated "plateaus" separated by "valleys" on the density map. The valleys are the clusters' boundaries. Therefore, a proper number of clusters can be determined.

There are some neurons "imaged" by certain input patterns in the feature map. According to the aforementioned feature map and density map, labeling the "imaged" neurons with the identity of the corresponding cluster forms a part of a discrimination map. There are still some blank neurons not labeled. A complete discrimination map can be obtained by applying the following equation to the blank neurons:

$$I = C(i) = \arg\min_j \|\mathbf{w}(i) - \mathbf{x}_j\|, \quad j = 1, 2, \ldots, p, \tag{2}$$

where I is the identity of a specific cluster, $C(\ )$ is a function indicating the cluster to which the blank neuron i belongs, $\mathbf{w}(i)$ is the weights of the neuron i, $\mathbf{x}$ is a input pattern, and p is the number of input patterns. The identity of a blank neuron is the identity of the input pattern which best matches the synaptic weights of the blank neuron. A complete discrimination map can then be obtained by labeling all blank neurons with their corresponding identities I as shown in Eq. (2).

3. Application and Discussions

The research area in this paper is the north area of Taiwan. About 34 rain gauges are available in the area. Among these gauges, gages 2, 6 and 31 are selected to validate the discrimination capability of SOMCD in this paper. The single-station design hyetographs of the 31 rain gages are obtained by analyzing the annual maximum events till 1998.[10]

The three maps (the feature, density, and discrimination maps) are derived using the SOMCD. In conventional cluster analysis, it is difficult to

determine an appropriate number of clusters. This problem can be easily
solved using SOMCD. The density map is adopted to help choosing a proper
number of clusters. The density map of single-station design hyetographs
is shown in Fig. 1. The void neurons are the valleys of the density map.
The bold lines drawn in the density map (Fig. 1) are the clusters boundaries. Because of these clusters boundaries, it is decided that the number
of clusters is four. The average single-station design hyetographs of the
four clusters are drawn in Fig. 2. Figure 2 demonstrates the homogeneities
within clusters and the heterogeneities among clusters. The clustering result
is quite reasonable.

The discrimination map is given in Fig. 3. The discrimination results
show that gages 2, 6 and 31 are allocated into the clusters C, A, and B,
respectively. The comparisons of the design hyetographs of the three validating gages with the average design hyetographs are shown in Fig. 4. The
results show that the allocations are quite reasonable.

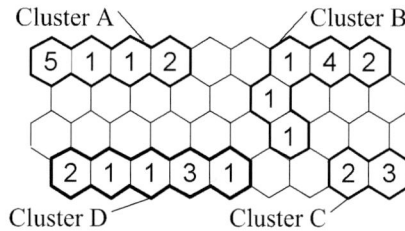

Fig. 1. The density map and the clusters' boundaries of single-station design
hyetographs.

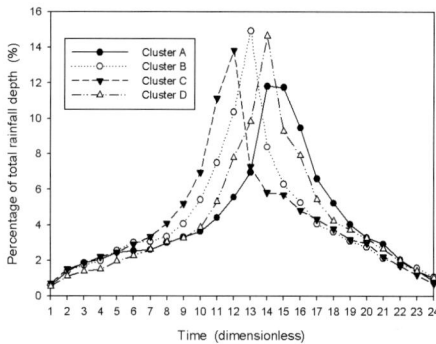

Fig. 2. The average design hyetographs of the four clusters.

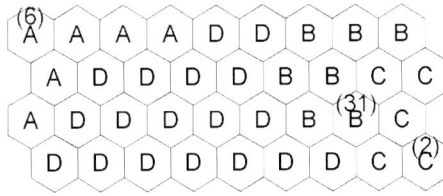

Fig. 3. The discrimination map and the clusters' boundaries of single-station design hyetographs.

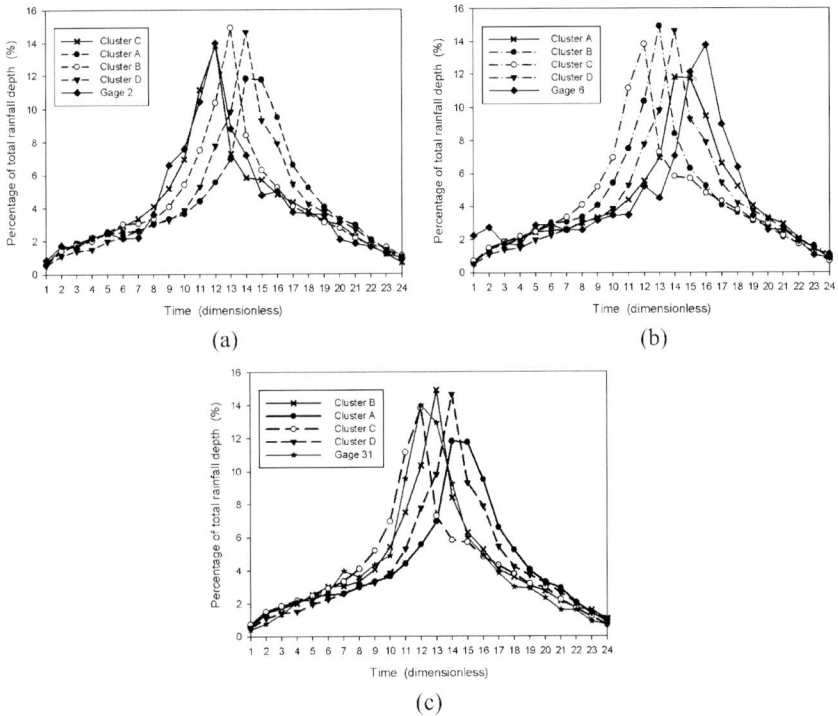

(a)

(b)

(c)

Fig. 4. Validations of the allocations of the three gages.

4. Conclusions

A method (SOMCD) that can combine the cluster analysis and discrimination analysis is proposed in this paper. A case study is performed using SOMCD to identify the homogeneities of the single-station design hyetographs in northern Taiwan. The assignments of validating gages to known clusters are also performed with the corresponding discrimination map. The results are both reasonable. It is concluded that the proposed SOMCD is an effective method for regionalization.

References

1. N. R. Bhaskar and C. A. O'Connor, *Journal of Water Resources Planning and Management* **115**, 6 (1989) 793–808.
2. R. J. Nathan and T. A. McMahon, *Journal of Hydrology* **121**, 217–238.
3. P. S. Yu, T. C. Yang and C. W. Liu, *Hydrological Processes* **16** (2002) 2017–2034.
4. M. P. Mosley, *Journal of Hydrology* **49** (1981) 173–192.
5. D. H. Burn, *Journal of Water Resources Planning and Management* **115**, 5 (1989) 567–582.
6. T. Kohonen, *Self-Organizing Maps* (Springer, New York, 1995).
7. P. Mangiameli, S. K. Chen and D. West, *European Journal of Operational Research* **93** (1996) 402–417.
8. S. C. Michaelides, C. S. Pattichis and G. Kleovoulou, *International Journal of Climatology* **21** (2001) 1401–1414.
9. S. Haykin, *Neural Networks: A Comprehensive Foundation* (Prentice-Hall, New Jersey, 1999).
10. K. S. Cheng, I. Hueter, E. C. Hsu and H. C. Yen, *Journal of the American Water Resources Association* **37**, 3 (2001a) 723–735.
11. K. S. Cheng, G. F. Lin, R. Y. Wang, M. H. Hsu, G. H. Yu, P. S. Yu and K. T. Lee, *Handbook for Hydrological Design*. Technical Report, Water Resources Agency, Ministry of Economic Affairs, Taipei, Taiwan (2001) (in Chinese).
12. R. A. Johnson and D. W. Wichern, *Applied Multivariate Statistical Analysis* (Prentice-Hall, New Jersey, 1992).
13. H. C. Riggs, *Investigations*, Vol. 4, Ch. B3 (US Geological Survey, Washington, DC, 1973).

FLOOD ANALYSIS AND MITIGATION ON LAKE ALBERT, UGANDA

ALBERT RUGUMAYO* and DAVID KAYONDO

Department of Civil Engineering, Makerere University, Kampala, Uganda
rugumayo@energy.go.ug

This study was carried out to provide knowledge for proper water resources planning, based on the statistical analysis and mathematical modeling. Major rivers contributing to the inflow and outflow of the lake were identified and their respective flow data acquired. Missing data were infilled using the normal ratio method and the ARMA model was used to extend the data. The annual maximum and minimum flow series were extracted from the data and modeled using the EV1 distribution. Because of its regular cross-sectional area, Lake Albert was also considered as a channel and therefore both reservoir and channel routing techniques were applied. The data sets considered were the maximum and minimum inflows, maximum and minimum outflows and an average year. Both the statistical modeling and flood routing provided useful results for flood modeling and mitigation.

1. Introduction

In 1961 and 1962, the flood plains of Lake Albert were greatly inundated to such an extent that Butiaba Inland Port was destroyed. The local inhabitants were forced to leave the Rift Valley floor, which constitute the lake flood plains and migrate to higher ground. Very little information, however, was available to planners on the trends of floods in this basin. The objectives of this study therefore, were to model Lake Albert's behavior during flooding using both statistical and mathematical methods and recommend appropriate flood mitigation.

Lake Albert as shown in Fig. 1 is found in western Uganda, along the Uganda — Democratic Republic of Congo (DRC) border. It is a rift valley lake, formed in the western arm of the Rift Valley, on the valley floor and is the second largest lake in Uganda after Lake Victoria. It is located in region $30°25'E$–$31°25'E$ longitudes, $1°N$–$2°15'N$ latitudes. Lake Albert covers an area of about $5{,}335\,\mathrm{km}^2$, is approximately $160\,\mathrm{km}$ long from tip to tip, $44\,\mathrm{km}$ wide at its widest width, $36\,\mathrm{km}$ at it's narrowest. At its deepest, the lake depth is about $51\,\mathrm{m}$. It is shown in Fig. 1. The catchment covers an area of $18{,}223\,\mathrm{km}^2$, which extends into the Democratic Republic of Congo

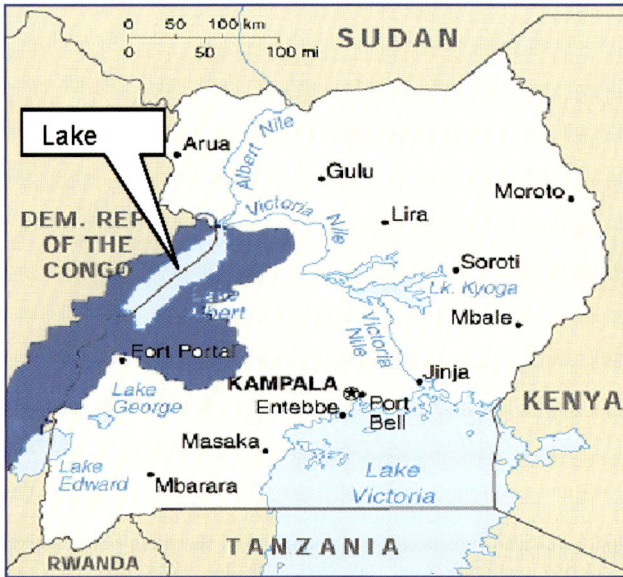

Fig. 1.   Uganda showing Lake Albert and its catchment area.

(DRC)[1]. The Lake Albert region is virtually the lowest point in Uganda at an altitude of 619 m above sea level, making it one of the hottest parts of Uganda. The geology primarily consists of banded gneisses of the basement complex rocks and therefore negligible seepage occurs.

The main inflow to the lake is contributed by rivers Semiliki, Muzizi, Nkussi, Wambabya, Waki I, Waki II, and Victoria Nile, while the outflow is through the Albert Nile. The total inflow to the lake is estimated by obtaining the sum of all the discharges of the above-mentioned rivers and these are the ones that were considered in the study.

2. Methodology

2.1. Data quality check

After identifying the rivers based on the map the catchment was delineated. Flow data for the rivers were obtained for the period 1950–2000 and a quality check was done using the double mass curve method.

The water balance for the lake was also obtained and is shown in Table 1.

Two methods were employed to infill missing data. These were the normal ratio for missing data in a period of less than 12 months and the

Table 1. Average annual water balance (1948–1970) for Lake Albert.

Parameter	Vol. ($10^6\,\mathrm{m}^3$)	Flow (m^3/s)	Total %
Rainfall on Lake	4	121	9
Inflow from Semiliki, other	3	147	11
Inflow from Kyoga Nile	30	954	73
Land catchment contribution	3	86	7
Total input	40	1308	100
Evaporation from the Lake	8	263	20
Outflow from Albert Nile	34	1070	82
Storage	−2	−25	−2
Total output	40	1308	100

Autogressive Moving Average ARMA (4) model[3] for missing data in a period more than 12 months. The reasons for using the ARMA model were firstly: because it gives the closest approximation to the regional series of flows; secondly, the model incorporates the variation in flows that are manifested as a result of the seasonality factor; and lastly previous success in applying these models on catchments in Uganda.[3]

2.2. *Frequency analysis*

From the data, the annual maximum and annual minimum series for both inflows and outflows were derived. The effective inflow of Lake Albert was obtained as the sum of all major inflows for the respective time intervals. The return periods for 2, 5, 10, 25, 50, 100, 150, and 200 years for the maximum inflow and outflow, minimum inflow and outflow and average inflow, were estimated by applying the method of moments. The year of average flow was determined by selecting the year whose annual maximum inflow was closest to the average of all the effective maximum inflows. It was assumed that these data series followed an Extreme Value Type One distribution given as

$$F(q) = e^{-e^{-((q-u)/\alpha)}}, \tag{1}$$

where q is the flow whose probability of occurrence is being determined $\alpha = (\sqrt{6}/\pi)\,\sigma$, and σ is the standard deviation and μ the mean of the data set of being analyzed $u = \mu + \alpha\gamma$, where γ is a constant and γ is 0.5772.

The return periods for each of these flows was then determined from the relationship:

$$T = \frac{1}{1 - F(q)}, \tag{2}$$

where T is the return period of the flow.

2.3. Flood routing[7]

Both Channel and Reservoir routing were performed for all the years of interest so as to determine Lake Albert's behavior either as a channel or reservoir in attenuating flood peaks.

2.3.1. Channel routing

In applying the technique of channel routing, the lake was assumed to be a channel through which water was being conveyed. This is justifiable as Lake Albert is oblong shaped, furthermore it is part of the western section of the Great East African Rift Valley.

The change in storage (channel storage) ΔS, was therefore first computed from the continuity equation given as

$$\bar{I} - \bar{O} = \frac{\Delta S}{\Delta t}, \tag{3}$$

where $\bar{I}$ is the average of inflows of two consecutive periods (months) $\bar{O}$ the average of outflows of two consecutive periods (months), and Δt is the time interval of one month. The storage volumes, S, were computed by summing the increments of storage from an arbitrary datum.

Values of S were then plotted against corresponding values of inflow and outflow.

It was assumed that the storage was a function of weighted inflow and outflow as given by the Muskingham equation as below:

$$S = K[xI + (1 - x)O], \tag{4}$$

where S, O and I are the corresponding values of storage, outflow, and inflow, respectively. K is the storage time constant for the reach and x is the dimensionless weighting factor between 0 and 0.5.

The values of x and K were obtained by trial and error x. Using the values of the storage S already obtained for the different years, a graph of S against computed values of $[xI + (1 - x)O]$ was plotted. Of the values of x assumed, the value that resulted in a graph conforming most closely to a straight line was selected as the x value representing the behavior of the lake. A line of best fit was inserted into the graph identified above using and the equation of the trend line obtain from the computer.

The inverse of the slope of the line of best fit computed as $(1/m)$ was obtained as the K value. The Muskingum routing relationship was then used to compute the values of the routed outflow O_2;

$$O_2 = C_0 I_2 + C_1 I_1 + C_2 O_1. \tag{5}$$

The K value obtained as the inverse of the slope "m" and the x values obtained were used to calculate the values of the Muskingham coefficients C_0, C_1, and C_2,:

$$
\begin{aligned}
C_0 &= \frac{-Kx + 0.5\Delta t}{K - Kx + 0.5\Delta t}, \\
C_1 &= \frac{Kx + 0.5\Delta t}{K - Kx + 0.5\Delta t}, \\
C_2 &= \frac{K - Kx - 0.5\Delta t}{K - Kx + 0.5\Delta t},
\end{aligned} \tag{6}
$$

K and Δt have the units of time. The theoretical value of K is the time required for an elemental (kinematic) wave to traverse the reach. It is approximately the time interval between the inflow and outflow peaks. Graphs of inflow, outflow, and routed outflow against time interval (months) were plotted for all the years of interest.

2.3.2. *Reservoir routing*

In reservoir routing, the Puls method was applied. The routing interval Δt was taken as 5 h (18,000 s). The outlet at Albert Nile was assumed to take on the behavior of a broad crested weir of which the outflow Q is given by the relationship:

$$Q = 1.7BH^{1.5}, \tag{7}$$

where B is the breadth of the weir and was measured as 4 km, which is the approximate width of Albert Nile at the outlet of Lake Albert. H is the arbitrary elevation of the surface of the water above the crest of the weir.

The volume, V, of water flowing out was computed as the product of the surface area of the lake obtained as approximately $5,335\,\text{km}^2$ and the elevation of the surface of the water above the crest of the weir, $H(\text{m})$. Arbitrary values of H were used ranging from 0.0 to 1.5 m increasing in intervals of 0.05 m. For all the years of interest, the flow was routed through Lake Albert in order to simulate its behavior. Similarly, the maximum and the minimum outflows were back routed through the reservoir. The change in variation of storage, over time for both inflows and outflows was determined and plotted.

2.4. Delineation of flood zones

This involved the demarcation of the levels of inundation of the flood plains for various return periods of the annual maximum inflows. The stage of flows of the selected return periods was determined by the use of the rating curves at the gauging station on the outlet of Lake Albert. A section of the lake and flood plains was then drawn and the water levels for the respective inflows of selected return periods indicated.

2.5. Field visits

The study area was visited, during which simple interviews were conducted concerning economic activities in the area, the flood history of the lake and other information, which was used to complement on the analytical findings of this study. Photographs of areas susceptible to flooding were also taken. Proposals for flood mitigation considered the feasibility, applicability, sustainability, affordability, and environmental impacts in the region.

3. Results

3.1. Statistical analysis

The statistical parameters for the selected data series were derived and are shown in Table 2.

The year of occurrence, the flow and return periods of the critical years are shown in Table 3. From the results, it is evident that the return period for the maximum outflow of (59 years) is greater than that of the maximum inflow (28 years). This implies that the maximum inflows into Lake Albert

Table 2.   Statistical parameters for the data series.

Data series	μ	σ	α	u
Annual max. inflows	2935.1	1935.0	1547.7	3828.4
Annual max. outflows	1341.1	465.69	363.10	1550.7
Annual min. inflows	1062.1	599.24	467.22	1331.8
Annual min. outflows	996.17	336.47	262.35	1114.7

Table 3.   Return periods of critical years.

Year	Year of occurrence	Discharge (m³/s)	$F(q)$	Return period
Maximum inflow	1964	8971	0.9646	28.24
Maximum outflow	1964	3029	0.9831	59.17
Minimum inflow	1974	1000	0.1308	1.150
Minimum outflow	1954	416	0.0000	1.000
Average	1969	2872	0.1565	1.190

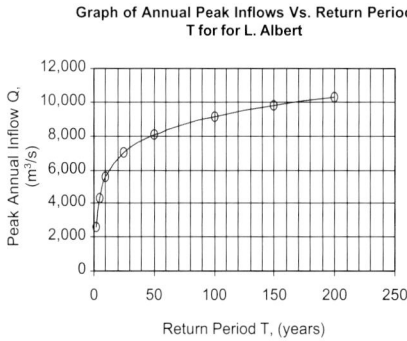

Fig. 2.   Graph of annual maximum inflow against return period.

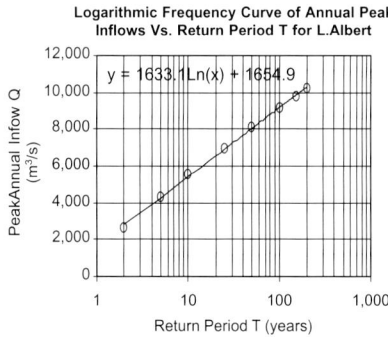

Fig. 3.   Semi-logarithmic plot of inflows against return periods.

occur more frequently than maximum outflows and the maximum inflow is nearly three times greater in magnitude to the maximum outflow. The plots of annual maximum inflows against return period are shown in Figs. 2 and 3. The plots for the outflows are similar in shape but with lower values.

3.2. Storage analysis

The results of the analysis are graphically presented in Figs. 4 and 5.

From the trends in the graphs above, it can be seen that the rate of charge of storage of the lake generally increases with increasing inflows and decreases with increasing outflows.

3.2.1. Channel routing

Trial graphs were then plotted of ΔS against S' for different time intervals for the values of x. The graphs of two values of x are as shown in Fig. 6.

Graph of change in storage ΔS (m³) Vs. Inflow of L. Albert (m³/s)

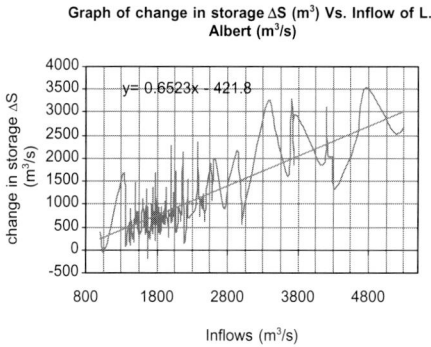

Fig. 4.   Variation of storage with inflow.

Graph of change in storage ΔS (m³) Vs. Outflow of L. Albert (m³/s)

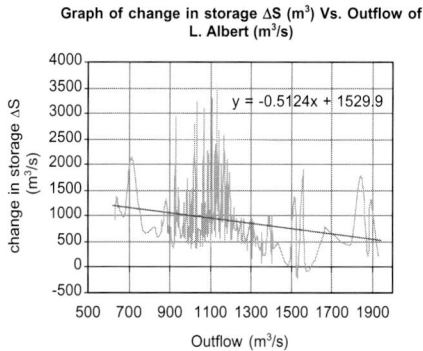

Fig. 5.   Variation of storage with outflow.

Graph of S' Vs. Storage ,S (x=0.25)

$y = 0.2474x + 1073.2$

S' = [x] + (1-x)O]

Storage S, (cumec month)

Graph of S' Vs. Storage ,S (x=0.4)

$y = 0.7862x + 562.74$

S' = [x] + (1-x)O]

Storage S, (cumec month)

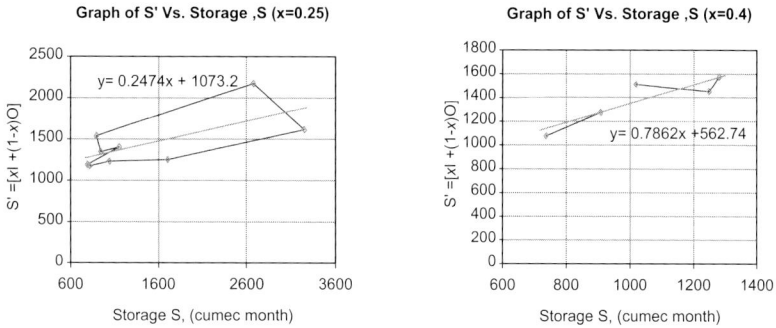

Fig. 6. Graphs of S' against S for determination of x and K for the channel reach.

As demonstrated in Fig. 6, the plot having $x = 0.4$ gives the best straight line. K was calculated as the inverse of the slope of the line of best fit. The gradient of the line was obtained as $S = 0.7862$ and the lag K is therefore taken as approximately 1.27 months, which is the time a flood wave would take moving from the inlet to the outlet of the lake. Muskingham coefficients C_0, C_1, and C_2 were estimated as -0.00695, 0.7986, and 0.2083, respectively.

The routed flows were then compared with the actual flows. From the channel routing graphs in Figs. 7–9, it is apparent that Lake Albert does not really approximate to a channel its behavior in attenuating flood peaks. In the case of Fig. 10, however, in the year of average inflow Lake Albert's behavior has been approximately simulated. This is significant.

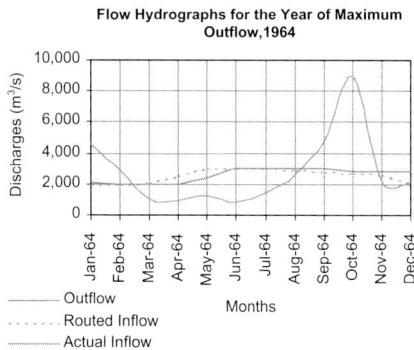

Flow Hydrographs for the Year of Maximum Outflow,1964

Discharges (m³/s)

Months

———— Outflow
· · · · · · Routed Inflow
———— Actual Inflow

Fig. 7. Graph of flows for the year maximum outflow (1964).

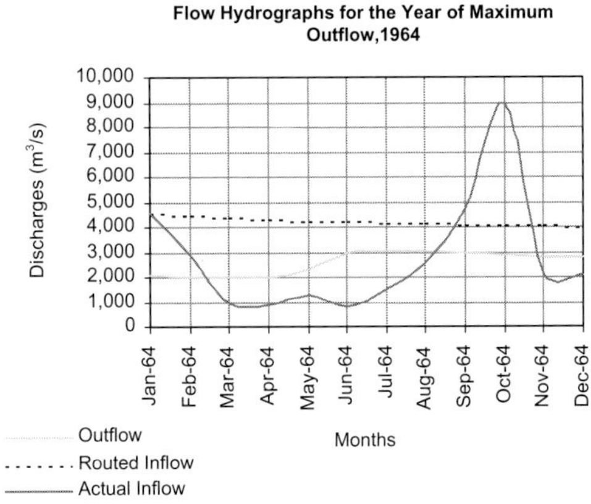

Flow Hydrographs for the Year of Maximum Outflow,1964

Fig. 8.   Graph of flows for the year of maximum inflow (1964).

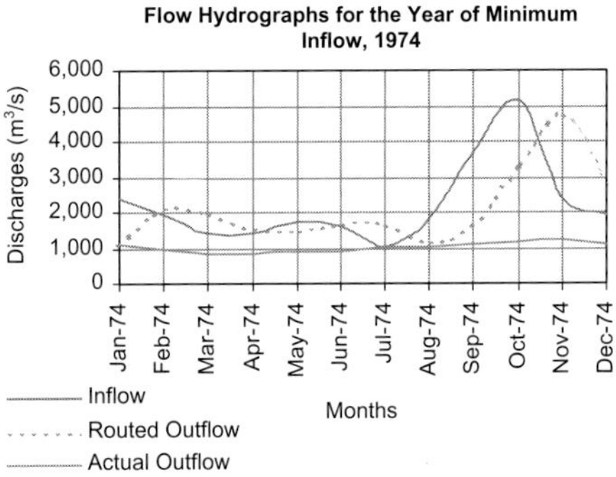

Flow Hydrographs for the Year of Minimum Inflow, 1974

Fig. 9.   Graph of flows for the minimum of year of average inflow (1969).

3.2.2. *Reservoir routing*

Figures 12–15 show that Lake Albert can be simulated using the reservoir routing model in attenuating flood peaks. A summary of the results of the flood routing is shown in Table 4.

3.3. *Flood plain delineation*

The demarcation of the flood zones is shown in Fig. 16, gives an indication of the susceptibility of the flood plain on the Uganda side of the lake, which is restricted to the Rift Valley floor.

Flow Hydrographs for the Year of Average Inflow, 1969

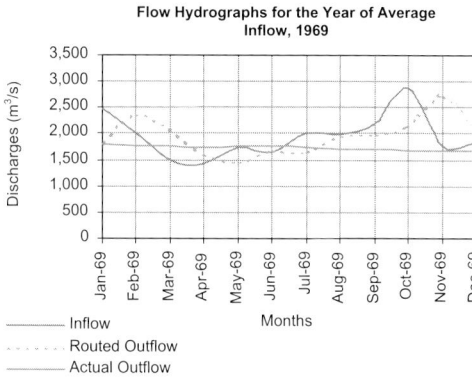

Inflow
Routed Outflow
Actual Outflow

Months

Fig. 10.   Graph of flows for the year of minimum inflow (1974).

Flow Hydrographs for the Year of Maximum Outflow,1964

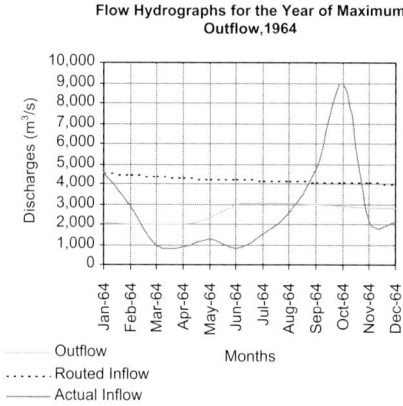

Outflow
Routed Inflow
Actual Inflow

Months

Fig. 11.   Reservoir back routing for the maximum outflow (1964).

4. Discussion

From the results obtained, it was established that Kyoga Nile contributes more than 70% of the inflow to Lake Albert. The implication is that Kyoga Nile is a key influence of the lake level. This is evidenced by the fact that the highest effective inflow to Lake Albert occurred at the same time as the peak discharge of Kyoga Nile. Hence the lake behavior can be predicted by observing Kyoga Nile discharges. The highest effective inflow of 8,971 cumecs from the data has a recurrence interval of about 28 years. On the other hand, the highest outflow of 3,029 cumecs had a return period

Flow Hydrographs for the Year of Maximum Inflow, 1964

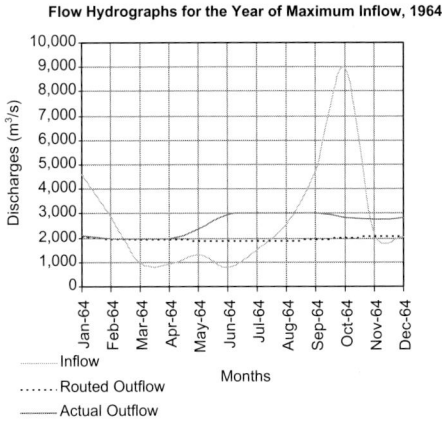

Fig. 12.   Reservoir routing for the year of maximum inflow (1964).

Flow Hydrographs for the Year of Minimum Inflow, 1974

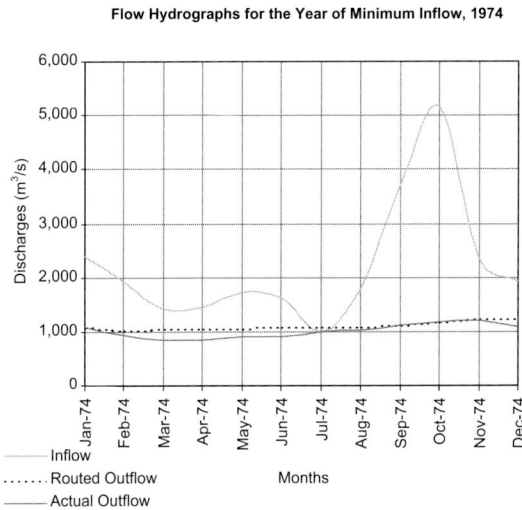

Fig. 13.   Reservoir routing for the year of minimum inflow (1974).

of 59 years. This result is similar to that obtained by a previous study on flooding on Lake Kyoga, which showed that the maximum inflow was about half of the recurrence interval for the maximum outflow, thereby contributing to flooding.[9] Furthermore, the maximum inflow is nearly three times the maximum outflow. This means that the maximum outflow would probably

FlowHydrographs for the Year of Average Inflow,1969

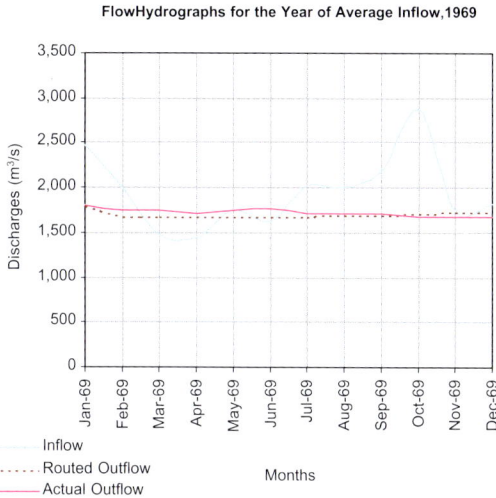

Inflow
Routed Outflow
Actual Outflow

Months

Fig. 14.   Reservoir routing for the year of average inflow (1969).

Table 4.   Summary of flood modeling of the Lake Albert.

Year	Channel routing	Reservoir routing
Maximum inflow, 1964	Poor	Excellent
Maximum outflow, 1964	Poor	Fairly good
Minimum inflow, 1974	Poor	Excellent
Average inflow, 1969	Fairly good	Excellent

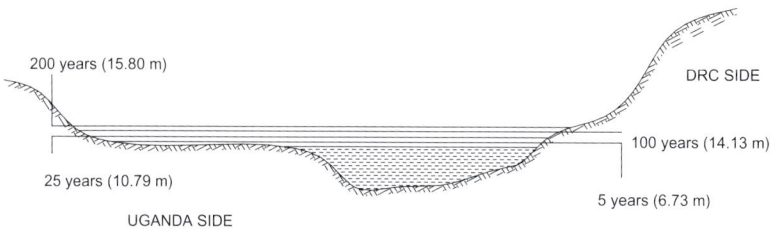

200 years (15.80 m)   DRC SIDE

                                                    100 years (14.13 m)
25 years (10.79 m)
                                          5 years (6.73 m)
     UGANDA SIDE

Fig. 15.   A cross section of Lake Albert showing levels of floods for various return periods.

occur once in a period during which the maximum inflow has been incident twice. This is presents a logical hypothesis that the lake floods periodically partly because the maximum inflow occurs more frequently the maximum outflow.

It was noted that the reservoir routing model produced very good results in simulating Lake Albert's behavior in attenuating flood peaks for the maximum, minimum and the average flows. The channel routing model, however, produced reasonable results in simulating Lake Albert's behavior for an average year. The results obtained from the process of delineation though acceptable, would have been more elaborate if an explicit and consistent empirical relationship had been established between the maximum effective inflows and the respective water levels at Butiaba. This would then cater for the variation in the cross-sectional area of the storage volume as the depth/water level of the lake increases in the computation of water levels of the lake.

5. Conclusions and Recommendations

Flooding in Lake Albert has been successfully modeled using both the frequency analysis and flood routing techniques.

The results from frequency analysis have been used to understand the behavior of the lake to develop flood demarcation zones in the flood plain. The reservoir routing model developed, provides a simple and accurate tool for flood forecasting.

The recommended measures for flood mitigation include establishing of an early warning system, sensitization of the local inhabitants, development of a land management policy establishment of a flood defences and installation of more automatic gauging stations.

Acknowledgments

The authors wish to express their gratitude to the Water Resources Management Department, Directorate of Water Development–Entebbe, for providing the hydrologic data that was used in carrying out this study.

References

1. Uganda Hydrological Network Map, Drawn and Published by Water Development Department 1974, Second Edition, Reproduced by the Department of Lands and Survey, Uganda, Reprinted 1997.
2. State of the Environment Report Uganda, 1996.
3. R. T. Clarke, FAO Consultant, Institute of Hydrology Wallingford U.K, "Mathematical Models in Hydrology FAO Irrigation and Drainage Paper 19" FAO 1973.

4. S. Muzira, Final Year Project Report — *Time Series Analysis on Eastern Uganda Catchments* (Makerere University Press, Kampala, 2000).
5. W. Viessman Jr. and G. L. Lewis, *Introduction to Hydrology*, Fourth Edition (Harper Collins College Publishers, New York, 1996).
6. E. S. Serrano, *Hydrology for Engineers, Geologists and Environmental Professionals* (Hydro Science Inc. Lexington, Kentucky, USA, 1997).
7. R. K. Linsley and J. B. Franzini, *Water Resources Engineering*, Third Edition (McGraw Hill, New York, 1979).
8. K. Brogan, ICE commission to review flood defence adequacy, *New Civil Engineer International* (March 2001), 42.
9. J. A. Makmot, Final Year Project Report — *Flood Mitigation on Lake Kyoga* (Makerere University Press, Kampala, 2000).

URBAN FLOOD ANALYSIS WITH UNDERGROUND SPACE

KEIICHI TODA*, KAZUYA INOUE and SHINJI AIHATA

Disaster Prevention Research Institute, Kyoto University
Gokasho Uji, Kyoto, Japan
*toda@taisui5.dpri.kyoto-u.ac.jp

In this study, an inundation simulation model based on a storage pond model is developed which can treat inundation of both ground and underground spaces in urban area. The continuity equation, momentum equation, and drop formula are used as the basic equations in the model. The model is applied to Fukuoka City in Japan, and Fukuoka flood inundation in 1999 is simulated in both ground and underground spaces. The computation results show good agreement with the actual record. The model is also applied to Kyoto City, Japan. The results show that the large inundation into underground malls and subways may occur in the case that the Kamo River overflows.

1. Introduction

When flood flow hits the central district of large cities, the inundation flow would extend to underground space and the damage would be serious. In fact, urban floods such as the ones that occurred in Fukuoka, Japan in 1999 and in 2003 and the one that occurred in Seoul, Korea in 2001 induced inundation into underground space and caused extensive damage. Therefore, it is very significant to study the inundation in underground space from the hydraulic and disaster preventive aspects.

Takahashi *et al.*[1] early made an inundation flow model in underground space. They treated the inflow from stairs into underground space as a stepped flow. They also showed that a horizontally two-dimensional (2D) inundation flow model can be applied to underground space. In their studies, only simple underground shapes were treated, and the effect of ceiling in underground space was not taken into account. The authors[2] developed the underground inundation model based on the one-dimensional (1D) network model with a slot, and applied it to Umeda underground mall in Osaka, Japan. They could succeed in treating the real complicated underground space and expressing both of the open channel flow condition and the pressurized flow condition. In their models, however, the flow behavior cannot be well simulated under some conditions of roughness coefficient, inflow discharge and slot area. In addition, many problems arose for acquisition and reduction of underground geographical data.

In view of these situations, a solid numerical simulation model named the storage pond model[3] is developed for underground inundation based on a pond model. This model is simpler than the above stated models. The data required for the model is reduced. Also, this model can treat both ground and underground inundation simultaneously without much difficulty. In this paper, the storage pond model is applied to the central area with underground spaces in two large cities in Japan, Fukuoka City and Kyoto City. Fukuoka City is the central one of western Japan, whose population is about 1.4 million, while Kyoto City is the old capital, whose population is about 1.5 million. The inundation flow behaviors in both ground and underground there are studied in detail.

2. Simulation Method

Underground spaces such as shopping mall are generally composed of stores, open spaces, subway entrances and basements of adjacent buildings. Although the total underground mall may be very complicated, it can be divided into some parts. If each part is assumed to be a storage pond that has its own volume, the underground space is expressed by the combination of storage ponds in the three dimensions (see Fig. 1). The slot is also incorporated in each pond taking into account the pressurized flow condition. Thus, the dispersion of inundation water can be expressed by obtaining the discharge flowing between the adjacent storage ponds. For the inundation on the ground surface, the similar method can be applied, without slot in each pond.

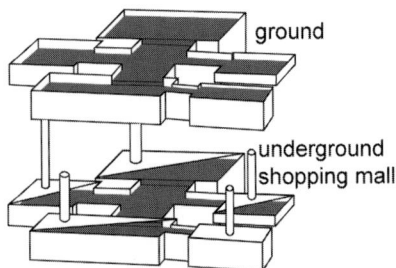

Fig. 1.   Storage pond model.

The governing equations are the continuity equation and the momentum equation expressed below as follows:

Continuity equation

$$A\frac{\mathrm{d}H}{\mathrm{d}t} = \sum Q_i + Q_{\text{ins}}; \quad A = A_f : \; h < D, \quad A = A_s : \; h \geq D,$$

where A is the effective base area of storage pond, H the water stage, Q_i the inflow discharge from ith adjacent storage pond, and Q_{ins} is the lateral inflow discharge from the ground surface. h is the water depth and D is the ceiling height of the storage pond. A_f is the area related to the storage pond shape and A_s is the slot area.

Momentum equation

$$\frac{L}{gA_b}\frac{\mathrm{d}Q}{\mathrm{d}t} = \Delta H - \alpha L Q \left|Q\right|,$$

where Q is the discharge, g the gravity acceleration, and L is the distance between the base area centroids of adjacent storage ponds. A_b is the cross-sectional area of adjacent storage ponds, and it is determined according to the water depths in the adjacent storage ponds. ΔH is the water level difference between the adjacent storage ponds and α is the loss coefficient associated with Manning coefficient.

At the inflow position from the ground surface to the underground space and the dropping position from the upper floor to the lower floor in multistory underground space, the following step flow formula is applied.

$$Q = B_e \mu_0 h_e \sqrt{gh_e},$$

where B_e is the effective width of entrance, μ_0 the discharge coefficient, and h_e is the water depth in the upper storage pond. If the lower storage pond is pressurized, then the above momentum equation is applied.

3. Application to Fukuoka City

3.1. *Studied area and computational conditions*

Figure 2 shows the studied ground area of about $2.8\,\mathrm{km}^2$ and Fig. 3 shows the ground elevation distribution. The ground elevation becomes lower from the Mikasa River to the direction of JR Hakata station. Figure 4 shows the underground space under JR Hakata station. (JR denotes Japan Railway.) The subway track space is assumed a storage pond with large volume. The total area and volume of underground space except subway railroad space are about $5.2 \times 10^4\,\mathrm{m}^2$ and $16.9 \times 10^4\,\mathrm{m}^3$, respectively.

K. Toda, K. Inoue and S. Aihata

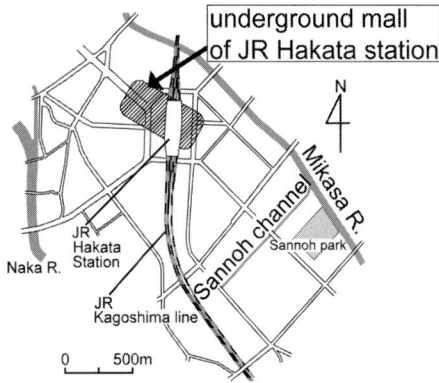

Fig. 2.   Studied area of Fukuoka City.

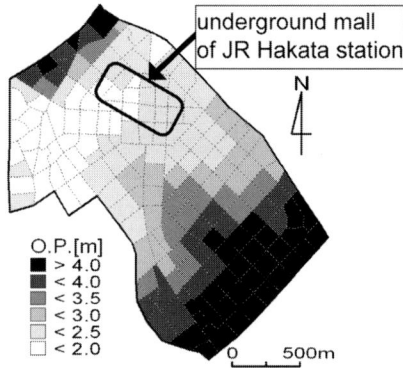

Fig. 3.   Fukuoka City ground elevation.

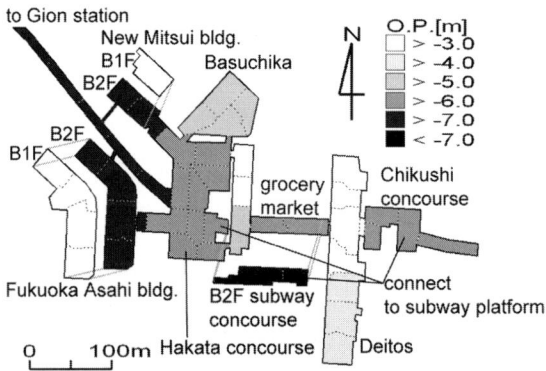

Fig. 4.   Studied area of JR Hakata station underground mall.

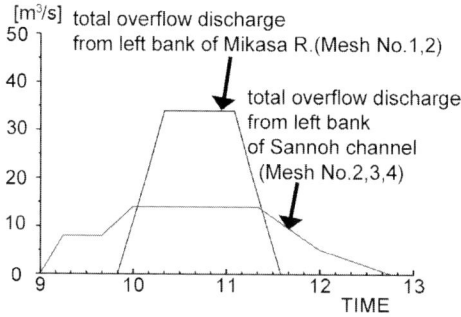

Fig. 5. Inflow discharge hydrographs.

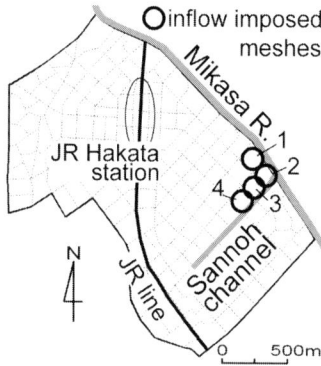

Fig. 6. Inflow imposed meshes.

As for the boundary condition, the overflow discharge from the Mikasa River obtained by Hashimoto *et al.*[4] is imposed. Figure 5 shows the discharge hydrograph and Fig. 6 shows the storage ponds to which the overflow discharge is allocated as the lateral inflow. The computation start time is 9:00 on June 29, 1999, when the overflow began from Sannoh channel.

In the computation, the drainage by sewerage system of 36.4 mm/h, 70% of the designed value is considered. The inflow water into the basements of buildings and the water stored in the storage tank of the underground mall are also considered. Steps of entrance to underground space and pavement (30 cm high in total) are also taken into account.

The values of Manning coefficient, n, are 0.067 for ground and 0.03 for underground, respectively, and the discharge coefficient of drop formula is 0.544. The computational time step Δt is set to be 0.05 s.

3.2. Computational results

Figure 7 shows the comparison between the computed maximum water depth distribution and the actual inundation record on the ground. The computed inundated area shows good agreement with the actual inundation area. By this computation, the inflow water volume into the basements of buildings and the water volume stored in the storage tank of the underground mall amount to 6.0×10^4 and $1.3 \times 10^4 \, \text{m}^3$, respectively.

Figure 8 shows the computed maximum water depth distribution and the actual inundation record of the studied underground space. The computed result is in almost good agreement with the actual record. The computed water volume flowing into the subway track space is about $5,000 \, \text{m}^3$

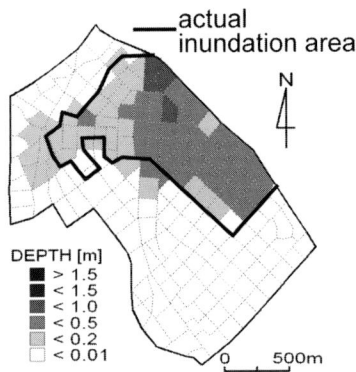

Fig. 7.   Maximum inundation depth on the ground.

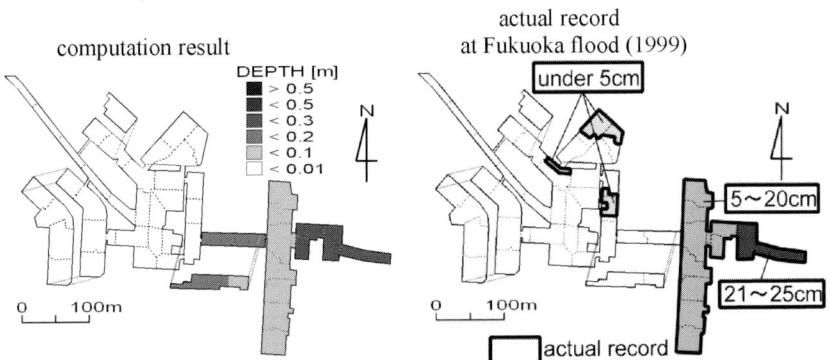

Fig. 8.   Maximum inundation depth of the underground.

at 13:00, while the actual water volume is assumed $1,000$–$2,000\,\mathrm{m}^3$. This difference may be caused by the accuracy of estimation of the water volume stored in the storage tank and the step height of the entrances to underground mall.

4. Application to Kyoto City

4.1. *Studied area and computational conditions*

Figure 9 shows the studied area of Kyoto City of about $51.6\,\mathrm{km}^2$. In the studied area, the underground spaces of Kyoto Oike underground mall, JR Kyoto station underground mall, the municipal subways and the railways privately operated are included. The Kamo River runs in the central area. Figure 10 shows the ground elevation distribution. Kyoto City is rather sloped from north to south.

In the computation, the river overflow in the central area is assumed and the hydrograph discharge shown in Fig. 11 is imposed at the mesh along

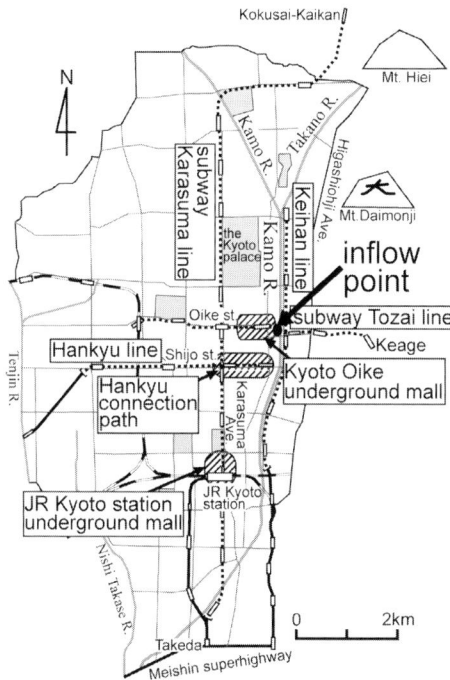

Fig. 9.   Studied area of Kyoto City.

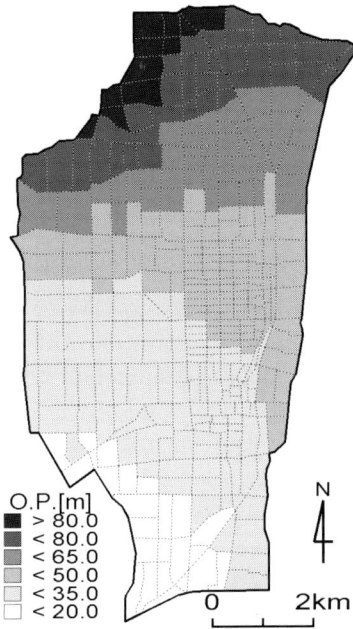

Fig. 10.   Kyoto City ground elevation.

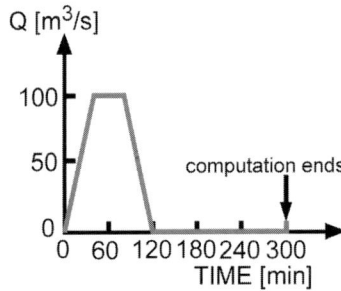

Fig. 11.   Inflow discharge hydrograph.

the river as the lateral inflow. The parameter values used here are the same as those in the preceding chapter.

4.2. Computational results

Figure 12 shows the computed inundation results of both ground and underground spaces at 90 min after. The inundation water disperses from the

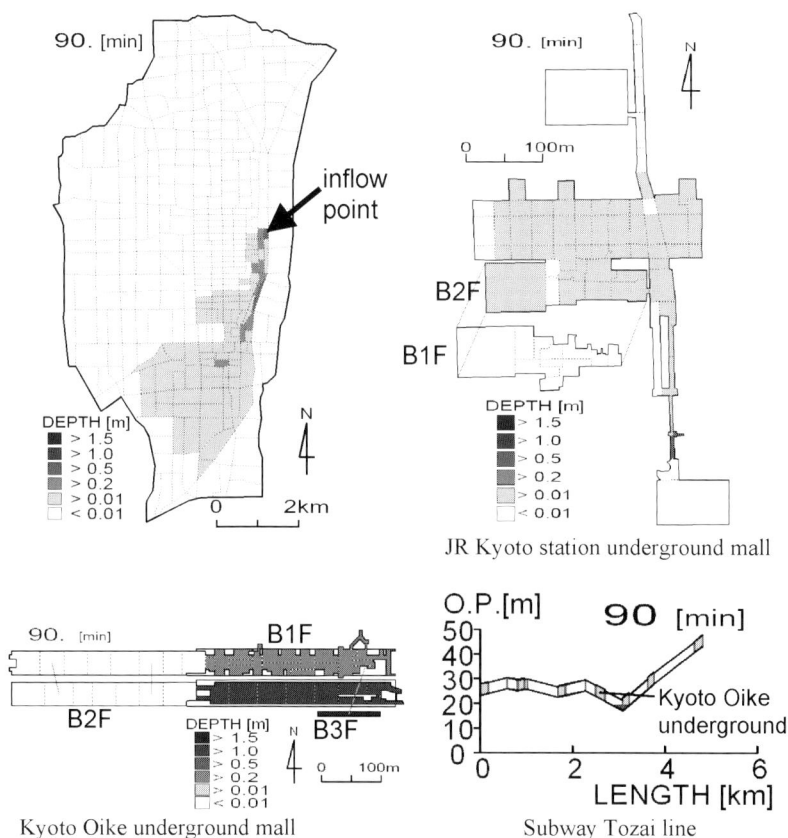

Fig. 12.   Computed inundation water depth distributions at 90 min.

overflow point to the south area, and it flows into Kyoto Oike underground mall, JR Kyoto station underground mall and subways. If the overflow occurs from this point, Kyoto Oike underground mall and the adjacent subway station are heavily submerged. In addition, the water flowing into the subway track space extends down to the next subway station. From these computational results, it is possible that the water flowing through subways inundates underground spaces that are far from inundation area of ground.

5. Conclusions

In this study, the simulation model is developed which can treat both of ground and underground water inundation in urban area. Through the application of this model, the risk of underground inundation by heavy rainfall or overflow from the river in a large city will be discussed in more practical way.

References

1. T. Takahashi, H. Nakagawa and I. Nomura, Simulation method on inundation in an underground space due to intrusion of overland flood flows, *The Annuals of the Disaster Prevention Research Institute*, Kyoto University, No. 33, B-2 (1990) 427–442 (in Japanese).
2. K. Toda, K. Inoue, O. Maeda and T. Tanino, Analysis of overland flood flow intrusion into underground space in urban area, *Journal of Hydroscience and Hydraulic Engineering*, JSCE **18**, 2 (2000) 43–54.
3. K. Toda, K. Kuriyama, R. Oyagi and K. Inoue, Inundation analysis of complicated underground space, *Journal of Hydroscience and Hydraulic Engineering*, JSCE **22**, 2 (2004) 47–58.
4. H. Hashimoto, K. Park and M. Watanabe, Overland flood flow around the JR Hakata-eki station from the Mikasa and Sanno-Channel River in Fukuoka City on June 29, 1999, *Journal of Japan Society for Natural Disaster Science*, JSNDS **21**, 4 (2003) 369–384 (in Japanese).

CLIMATIC VARIABILITY AND DROUGHT IN RAJASTHAN

ABHINEETY GOEL* and R. B. SINGH[†]

Department of Geography, Delhi School of Economics
University of Delhi, Delhi — 110007, India
*abhineetygoel@yahoo.co.in
[†]rbsgeo@hotmail.com

In 2002, the Government of India acknowledged severe drought and scarcity in all 32 districts of Rajasthan. It recognized large number of deaths in Baran district as a result of this situation. This paper presents a case study of severe drought in Baran district, as resulting from climatic variability in southeast part of Rajasthan. It further evaluates the post-drought impact on the livelihood security of Saharia tribe, Rajasthan's only primitive tribe.

1. Introduction

As the elusive southwest monsoon played truant around India, many states have faced a gloomy harvest and a year of food shortages and drought. Rajasthan, one of the western states of India, faces the problem of drought more frequently than any part of the world. It has faced drought for 44 of the last 50 years. The arid climatic condition and ever degrading land use practices are the major culprits behind these recurring phenomena of drought. Loss of lives and property and subsequent out-migration is the result and thereby putting pressure on the economic condition of the state. A major drought affected Rajasthan in 1987–1988, when the country faced the worst drought of the century. A decade later, the state suffered severe drought conditions for three successive years: 1998–1999, 1999–2000 and 2000–2001. 2002 was the fifth consequent year of drought. In 2002, 32, districts, a total of 41,711 villages and 45 millions of population were affected.[1] Although the whole state was affected by the drought, some districts in western Rajasthan, like Barmer, Jaisalmer, and Jodhpur were hit severely because of the fragile agro-ecology of the region. But surprisingly, Baran district, situated in the southeast, which experiences abundant rainfall, was also adversely affected by the drought. The present paper is an attempt to identify the driving factors and assess the impact of climatic aspects of this disastrous phenomenon.

2. Study Area

Rajasthan, with a total geographical area of 342,239 square kilometers and total population over 56 millions (2001), is India's largest state. The state is characterized by a non-nucleated, dispersed pattern of settlement, with diverse physiography ranging from desert and semiarid regions of western Rajasthan to the greener belt east of the Aravallis, and the hilly tribal tracts in the southeast. In Rajasthan, winter temperature ranges from 8°C to 28°C and summer temperature ranges from 25°C to 46°C. Average rainfall in Rajasthan also varies from about 101.60 to 660.40 mm annually. Considering the physical as well as climatic characteristics of southeastern part and its frequent experience with drought in the last few years, Baran district has been selected as study region. Baran district was carved out of Kota district in 1991 (Fig. 1). It extends from 24°25'N to 25°25'N and 76°12'E to 77°26'E. Known as the "Green Belt of Rajasthan," it has total forest area of 0.21 million hectares. Its total population is 1,021,653 (2001). Baran has dry climate except in monsoon season. Main soil is black *kachari* soil, which is highly fertile, and stony soil is found in southern and eastern parts.[2]

3. Methodology

The present study is basically based on analysis of data derived from various sources. From the data, it has been tried to establish relationship between rainfall and drought conditions. Since drought condition hit people of the region, in terms of less availability of food grains and drinking water, the study has attempted to analyze the relationship between rainfall and agricultural production.

4. Droughts and Livelihood Security in Rajasthan

More than 60% of the Rajasthan's total area is desert, with sparsely distributed population, entailing a very high unit cost of providing basic services. It has a predominantly agrarian society, with nearly 70% of its population depending on agriculture and allied activities. However, nearly 80% of all land in the western desert districts is unfit for farming. Agriculture continues to be dependent on rainfall. Failure of the monsoon causes

Location of Baran district in Rajasthan, India

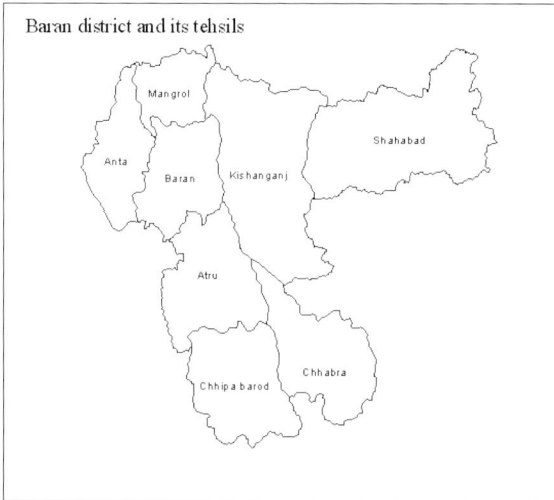

Fig. 1.   Location of study area.

severe drought and scarcity conditions. Rajasthan has only 1% of India's total water resources, and irrigation covers about 30% of the total cropped area.[3] The economy of Rajasthan is characterized by diversity in terms of livelihood sources and low levels of income, poverty, and unemployment. Rajasthan is deficient in water (surface and ground). Ground water at many places is unfit for human and livestock consumption. Socio-economic infrastructure is also deficient. Low productivity of agriculture and the dimension

of ecological risk make food security and subsistence the primary concern
of farmers. Set within this diverse geographical terrain, Rajasthan encom-
passes a wide range of livelihoods.

5. Major Driving Forces for Drought

Droughts are a recurring phenomenon. The India Meteorological Department defines a failure of the monsoon as a year in which the actual rainfall has been 20% less than the "normal" rainfall. By this definition, in
Rajasthan the monsoon failed in just two years — 1987 and 2002.

The average annual rainfall in Rajasthan is 531 mm; compared to the
all-India average of 1,100 mm. Rainfall is not only low, but also uncertain.
Further, there are wide variations in rainfall across the state — from as
little as 193 mm in Bikaner in the west, to as much as 607 mm in the eastern
plains in normal years.

The state received just 220.4 mm rainfall up to September 30, 2002,
against the normal of 518.6 mm in the overall monsoon. The overall state
rainfall deficit was –57.5%. Overall in 2002, the monsoon was worst for
Rajasthan in last 17 years as the state recorded a minimum of 220.4 mm
rainfall against the normal of 533 mm. A maximum average rainfall of
726 mm was recorded in 1996 and minimum 291.6 mm was recorded in 1987
but precipitation went down to 220.4 mm as on September 30, 2002.[1]

According to the analysis of rainfall received during monsoon period
(from June 1, 1999 to September 30, 1999), the districts of the State have
been categorized into six different categories (Table 1).

The state Government felt that although 15 districts had received deficit
rainfall within normal limits of rainfall, as they lay between (+) 20%, but
because of long dry spells, crops got damaged. Although for the state as
a whole, the overall deficiency was only 16%, this deficiency varied from
district to district and the dry spells created havoc. Although rains were
sufficient in terms of quantity in the month of July 1999 but the rains
occurred only in the last week of July 1999. As such sowing of Kharif crops,
particularly cereals, got delayed which affected the production prospects
adversely.[4]

As per estimates of the Government of Rajasthan, 23,406 villages out
of a total 34,693 villages of 26 affected districts (total number of districts
is 32) were affected by drought in the year 1999–2000. The loss of the
crops ranged from 75% to 100% in 18,085 villages, whereas loss occurred
from 50% to 74% in 5,321 villages. The human population affected in these

Table 1.   Districts under different categories of rainfall: 1999.

Category	Name of districts	Its number
1. Abnormal rainfall (60% or more than normal rainfall)	Nil	Nil
2. Excess rainfall (20%–60% more than normal rainfall)	Jaisalmer, Bharatpur, Chttaurgarh	3
3. Normal rainfall (−20% to +20% of normal rainfall)	Hanumangarh, Jodhpur, Ajmer, Bhilwara, Nagaur, Tonk, Dausa, Dholpur, Jhunjhunu, Sikar, Baran, Bundi, Jhalawar, Karauli, sawai Madhopur	15
4. Deficit rainfall (−60% to −20% less than normal rainfall)	Bikaner, Churu, Ganganagar, Jalore, Sirohi, Pali, Alwar, Jaipur, Kota, Banaswara, Dungarpur, Rajsamand, Udaipur	13
5. Scanty rainfall (−60% or less than normal rainfall)	Barmer	1

Source: Ref. 4.

villages was 261.79×10^5. It is said to be the second drought year, as last year also, in all 20,069 villages of 20 districts of the State were affected by scarcity.

With good rains in the month of July 2000, there was every hope that the crisis would be over. However, it is distressing that the widespread failure of rains in the month of August and September 2000 has raised the spectre of a much more serious drought situation, since this was the third consecutive year of rain failure in Rajasthan. While the overall deficiency in rainfall from June to September was 29%, 21 districts out of 32 districts had rainfall deficiency to the extent of (−) 60% to (−) 20% (Table 2).

6. Impact of Drought in Rajasthan

In 2002, out of the 5.87 million hectares of area sown, 2.72 million hectares had been damaged due to scarcity of rainfall. The western districts of Rajasthan were most affected by the unavailability of fodder from the neighboring States (Punjab and Haryana), which used to supply fodder in normal years. Due to non-availability of water and fodder, people started leaving their unproductive cattle and there were reports of animals dying of hunger and thirst.

Table 2.   Districts under different categories of rainfall: 2000.

Category	Name of districts	Its number
1. Abnormal rainfall (60% or more than normal rainfall)	None	Nil
2. Excess rainfall (20%–60% more than normal rainfall)	Nagaur	1
3. Normal rainfall (−20% to +20% of normal rainfall)	Barmer, Jaisalmer, Jalore, Dausa, Baran, Bundi, Karauli, Kota	8
4. Deficit rainfall (−60% to 20% less than normal rainfall)	Bikaner, Churu, Ganganagar, Hanumangarh, Jodhpur, Pali, Sirohi, Ajmer, Bhilwara, Tonk, Alwar, Bharatpur, Dholpur, Jaipur, Jhunjhunu, Jhalawar, Sawai Madhopur, Banaswara, Chittorgarh, Dungarpur, Udaipur	21
5. Scanty rainfall (−60% or less than normal rainfall)	Sikar, Rajasamand	2

Source: Ref. 4.

The Kharif crops were adversely affected by drought during last three years (1997–1999) in general, and during last two years in particular. Although the rainy season crops were sown in large areas, the crops failed in most of the sown area during 1998 and 1999. The effect of drought was more pronounced on fodder availability and the increase in price of fodder was more pronounced (2–3-fold).

In normal years, most farmers in the region prefer to take rainfed crops during kharif season and irrigated crops during rabi season. This practice is followed because of the fact that the groundwater is mostly saline. Taking rainfed crops during rainy season helps in reduction of salt load in soil profile due to leaching with good quality rainwater. But acute shortage of fodder, and to some extent food also, forced them to take irrigated kharif crops, particularly the pearl millet.

The effect of drought on livestock was mostly due to drastic scarcity of water, fodder and feed. Roughage stock of farmers was almost exhausted during 1998. Most of the cattle were let loose on their own, leading to their mortality in some cases. The mortality rate was high in cattle and sheep. The possible causes for livestock mortality during drought were: malnutrition and debilitated weak conditions, negative protein balance caused by wheat straw, intake of highly saline/contaminated water, and incidence of various diseases. Livestock migration is a predominant characteristic of arid and semiarid areas of Rajasthan. It was found that 78% of the livestock migrated from the Barmer district followed by 70% from the Jaisalmer and 20% from the Jodhpur district.

7. Scenario of Drought in Baran District

In 34 years, the rainfall in Baran significantly varied with a low of 26 mm to as high as 131 mm (Fig. 2). On the other hand, there was not much significant variation in temperature. The inhabitants of Baran are engaged in agricultural activities (producing crops like wheat, pulses, maize, sugarcane amongst others), growing livestock, and industries. There are a total of 119 agricultural based industries.[5] Livestock contribute more than 50% of the GDP in these areas (Table 3).

Rainfall in Baran (1970-2004)

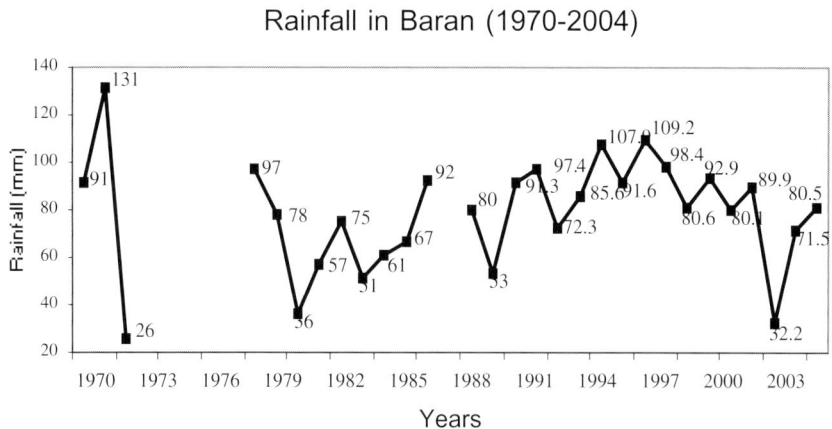

Fig. 2.   Rainfall in Baran district (1970–2004).[15,16]

Table 3.   Livestock in Baran district.

Types	Numbers
Cattle	420685
Buffaloes	103513
Sheep	16041
Goats	125962
Camels	2881
Pigs	5432
Horses and ponies	3474
Mules and donkey	2387

Source: Ref. 6.

Other factors include inadequate water availability, high water loss in storage and distribution, utilities, over exploitation of surface and ground water, shift in agricultural practices (low-to-moderate water demanding crops to high demanding crops), crop damage due to erratic rain and pest and last but not the least, rapid growth rate of population and animals.

8. Relationship between Rainfall and Agricultural Production

Rainfall and agricultural operations in Rajasthan, and indeed in many parts of India, are very closely related. Therefore, rainfall and land area sown tend to be positively correlated. In Rajasthan, only 25% of agricultural land is irrigated. The area sown is slightly better indicator of a drought because it reflects one of the ways in which lack of rainfall affects the human life. Most agriculture in Rajasthan is rainfed, and people's livelihoods are still quite heavily dependent on agriculture. According to Department of Agriculture, the production of Rabi crops in Baran district till April 30, 2003 was only 40%, whereas usually it is 80%–90%. During *Kharif* season, the sowing of crops was done in 5.87 million hectares against the targeted 12.9 million hectares, i.e. 45.5% of the normal sowing (Fig. 3). Out of the 5.87 million hectares of area sown, 2.72 million hectares has been damaged due to scarcity of rainfall.[7]

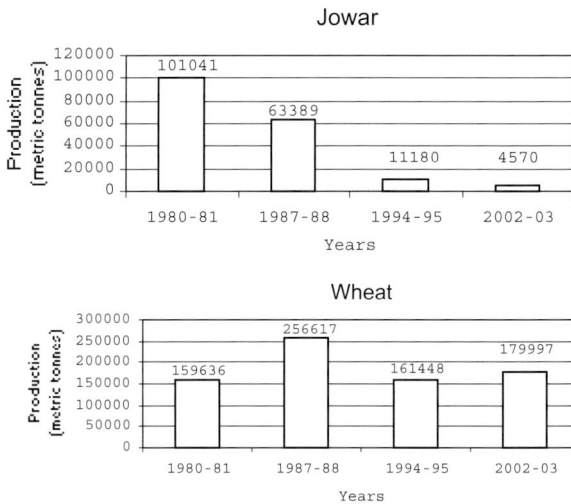

Jowar

Wheat

Fig. 3.   Production of two major crops in Baran (1980–2003).[13,14]

Table 4.   Fortnightly rainfall position in Baran district: 2000.

	June 1 to June 30	July 1 to July 15	July 16 to July 31	Aug. 1 to Aug. 15	Aug. 16 to Aug. 31	Sept. 1 to Sept. 15
Normal (in mm.)	81	158	158	144	154	63
Actual (in mm.)	3	101	409	83	103	162
Percent deviation	96	36	159	42	33	157

Source: Ref. 4.

The deviation of the rainfall from the normal was significantly experienced in Baran district during 2000 (Table 4).

All the five towns of Baran district have already been covered under drinking water supply scheme and there are 39 water supply schemes operating there. In 2002–2003, it had 241,640 hectares of net irrigated area and 26,477 hectares of total irrigated area. The ground water level went down causing the drying of about 30% of the hand pumps. Poor maintenance of water harvesting structures like *talab* and *talais* and indiscriminate exploitation of groundwater by big rich farmers further worsened the situation. Due to the scarcity of water sources, and to the high cost transportation to western Rajasthan, water rates rose. Increased cultivation, in turn, had put tremendous pressure on groundwater resources. Due to malnutrition and infections, the price of sheep and goat fell from Rs 500–Rs 1,000 to Rs 200–250.[8]

In Baran district, there is only one district government hospital, 201 subhealth centers, and 32 primary health centers. There are only four electrified municipal towns and 1,162 electrified villages.[2] Sahariya, the only tribal community, faced the consequences of drought most severely. Most socially and economically marginalized people, they had to face the scarcity of both drinking water and food grains. In some of the regions these tribal people were forced to eat the local grass — *Saawa*, these grasses are poisonous in nature if consumed in more quantity. Due to the far off locations, even government support in terms of water tanker and food depot could not reach in time. Availability of medical facilities was also disrupted due to the same reason.[9]

9. Findings

Meteorological factors mainly caused drought in Baran district. Rainfall and land area sown tend to be positively correlated. It has been found that

the productivity of the major crops declined significantly with decrease in
annual rainfall and consequently the region has experienced the food and
fodder scarcity. Some of the deaths are also reported in the region not only
due to the extreme climatic phenomena but also from the poor infrastruc-
tural conditions in terms of non-availability of good and metalled approach
road, far-off location of the medical facilities and not proper arrangement
of the drinking water supply through tankers and other means.

10. Conclusion

Considering drought as a regular feature of the economy, its timely identification should be focused so that they do not develop into a crisis. Further, following steps should be encouraged like judicious use of irrigation
water, rainwater harvesting, management of underground water, improved
agricultural practices, diversification of livelihoods and finally people's par-
ticipation should be encouraged.

Acknowledgments

We would like to thank the library staff at IARI, Pusa and India Meteorological Department for helping us to collect data for Rajasthan.

References

1. Rajasthan Drought Situation Report, 2002, Government of Rajasthan.
2. http://www.baran.nic.in/dprofile2.htm
3. Rajasthan Human Development Report 2002, published by Government of
   Rajasthan.
4. L. C. Gupta and V. K. Sharma, Drought in Rajasthan, National Centre for
   Disaster Management, Indian Institute of Public Administration, New Delhi,
   2001.
5. http://www.baran.nic.in/industries2.htm
6. *Akaal Samiksha* Report, Report of center of community economics and development consultants society, KBS printers, Jaipur, 2002.
7. http://www.baran.nic.in/statistics2.htm
8. R. Khera, Discussion Paper Series — 5 Drought Proofing in Rajasthan:
   Imperatives, Experience and Prospects, Human Development Resource Center, UNDP India, 2003.
9. N. Mishra, Hunger deaths in Baran, *Frontline* **19**, 24 (2002).
10. M. Bokil, Drought in Rajasthan, in search of a perspective, *Economic and
    Political Weekly*, November 25, 2000.

11. Directorate of Economics and Statistics, *Indian Agricultural Statistics*, Vol. II, District Wise (Department of Agriculture, Ministry of Agriculture, 1987–1988).
12. Directorate of Economics and Statistics, *Indian Agricultural Statistics*, Vols. I and II, All India State-Wise and District-Wise (Department of Agriculture, Ministry of Agriculture, 1992–1993).
13. Indian Meteorological Department, Climate of Rajasthan State, Government of India, Indian Meteorological Department, Delhi, 1988, pp. 147–152.
14. Indian Meteorological Department, Indian Daily Weather Report 1970–2004, Indian Meteorological Department, Delhi, 2004.

SIMULATING SCENARIO FLOODS FOR HAZARD ASSESSMENT ON THE LOWER BICOL FLOODPLAIN, THE PHILIPPINES

MUHIBUDDIN BIN USAMAH and DINAND ALKEMA*

Int. Institute for Geo-Information Science and Earth Observation (ITC)
PO Box 6, 7500 AA Enschede, The Netherlands
alkema@itc.nl

This paper describes the first results from a study to the behavior of floods in the lower Bicol area, the Philippines. A 1D2D dynamic hydraulic model was applied to simulate a set of scenario floods through the complex topography of the city Naga and surrounding area. The simulation results are integrated into a multi-parameter hazard zonation for the five scenario floods.

Introduction

The Bicol floodplain is located on Luzon Island, the Philippines (Fig. 1). Floods are the most serious natural hazard in this area. They occur almost every year and cause substantial suffering, loss of life and economic damage. This is especially true for the city of Naga, that is located at confluence of the Bicol river with the Naga river. Simulation of flood events in this region using a two-dimensional (2D) flood model can contribute to minimizing the loss of life and damages to property and can be a helpful support for flood risk management.

Flood Modeling

This research applies a 1D2D flood propagation model SOBEK, in the Lower Bicol Floodplain to simulate flood events with a return period of 2, 5, 10, 25, and 50 years. In combined 1D2D flood modeling, the flow of water within the riverbanks, is modeled as one-dimensional (1D) flow. At the moment that overland flow occurs (because one of the banks is over-topped), the flow is computed using a 2D solution of the flow equations (Fig. 2). SOBEK is a grid based inundation model that uses the finite difference method. The calculation scheme is based upon the following

M. B. Usamah and D. Alkema

Fig. 1.   Location of the study area.

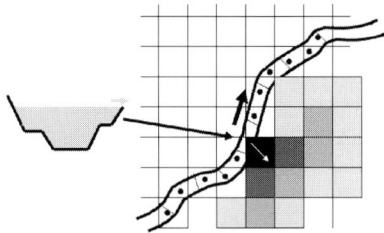

Fig. 2.   Coupled 1D-2D modeling.

characteristics[1]:

- The continuity equation is approximated such that (a) mass is conserved not only globally but also locally and (b) the total water depth is guaranteed to be always positive which excludes the necessity of "flooding and drying" procedures.
- The momentum equation is approximated such that a proper momentum balance is fulfilled near large gradients.

The combination of positive water depths and mass conservation assures a stable numerical solution that converges thanks to the momentum balance. The numerical schemes, described in depth by Hesselink *et al.*,[1] are tuned for speed, needed to assess every thinkable flooding scenario.[2] Both the progressive wave phase and the basin filling can be described accurately so that SOBEK can tackle flow over initially dry land and flow phenomena occurring shortly after a dike break. Since SOBEK computes on a rectangular grid and geometrical input data can be specified in a number of ways, land layout features, as dikes, roads, railroads, waterways, viaducts, etc. can easily be included in the analysis. The user can force dike failures so that "what if" scenarios can be investigated. A more elaborate explanation of this coupled dynamic hydraulic is given, e.g. in Refs. 1–3.

Model Requirements

The 1D2D flood simulation tool SOBEK requires four input parameters: a Digital Elevation Model (DEM), surface roughness coefficients of the floodplain, the river network with cross-sections and hydrological input data. The DEM was generated from interpolation of spot heights and contour lines obtained from different sources and scales (see Fig. 3). The hydrological data consisted of discharge and water level time series derived from a MIKE-11 Model by Nippon Koei[4] and NEDA,[5] for the upper boundaries (Bicol, Naga, and Libmanan River) and water level at the lower boundary (San Miguel Bay) for the return periods of 1.25, 5, 10, and 25 years. Values for return period of 2 and 50 years are derived through interpolation and extrapolation of the available data. The software generates maps of water depth and flow velocity at predefined intervals, as well as time series of certain parameters (water level, discharge, etc.) at predefined locations.

Results

Figure 4 shows the maximum water depth and flow velocity for the scenario flood with a return period of 10 years. The output maps can also be used to

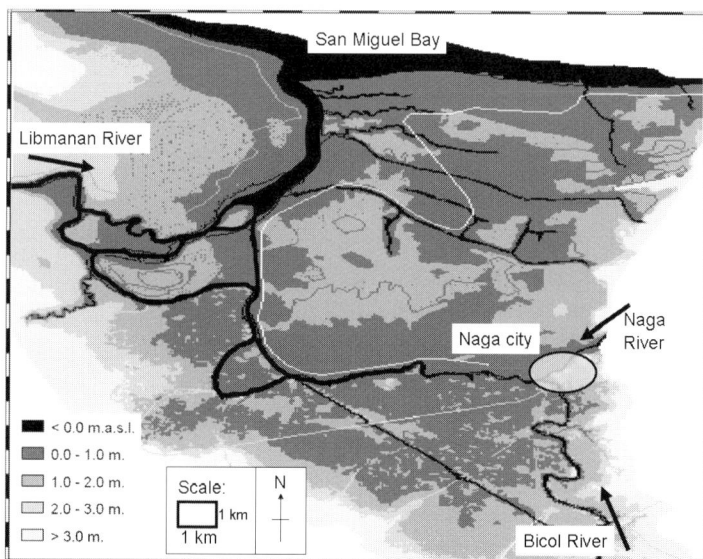

Fig. 3. Overview of the study area; scale varies in this perspective view. Elevation data are in meters above sea level (m.a.s.l.).

Maximum water depth (10 year flood)   Maximum flow velocity (10 year flood)

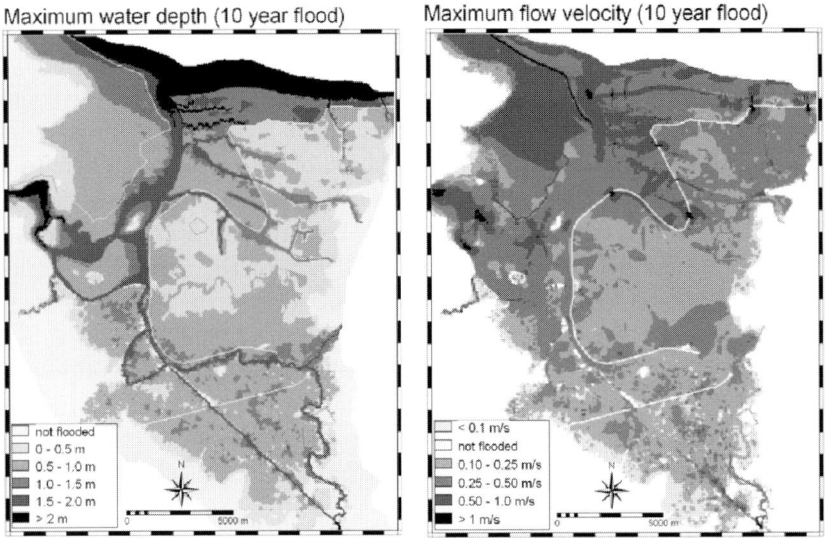

Fig. 4.   Result maps of the modeling (10 years flood scenario); Left: maximum water depth. Right: maximum flow velocity.

generate additional parameter maps that could be of use for the assessment of damage and risk, for instance maps showing the propagation of the flood, the kinetic energy of the flood water, the estimated duration, etc. The model shows an increase in the inundation extent of about 50% from flood with a return period of 2 years to the largest flood with a return period of 50 years.

Multi-parameter hazard assessment

Indicator maps produced from modeling were further processed to create flood hazard maps. Understanding that the level of flood hazard cannot be measured by single parameter, a hazard categorization was made based on work from Penning-Rowsell and Tunstall,[6] who proposed three hazard categories based on a combination of flow velocity and water depth (Fig. 5). The resulted hazard zonation for five return periods is shown in Fig. 6

Conclusions

A combined 1D2D model is an ideal tool to simulate riverine floods in low-land and river delta river areas. It combines the strengths of 1D catchment models that simulates flow within the riverbed with the possibility to study

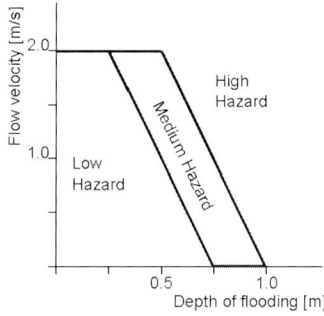

Fig. 5.   The categorization of flood hazard based on velocity and water depth.[6]

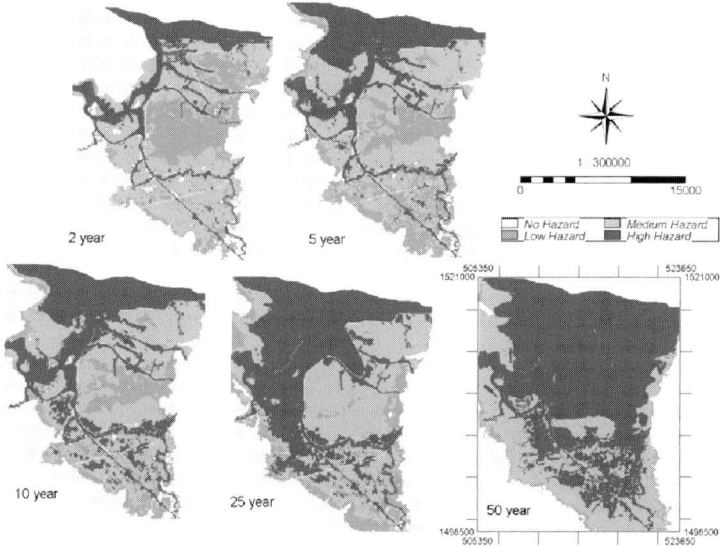

Fig. 6.   Flood hazard map: combination of water depth and flow velocity according to Penning-Rowsell and Tunstall.[6]

in greater detail the propagation characteristics when the floodwater disperses and flows over complex topography, like after overtopping or breaking through the levees, or in an urban environment. An advantage of 2D modeling is that it gives information on the kinetic energy (impulse/momentum) of the flood water and thus allows the assessment of flood propagation characteristics (like warning time). This results in a more realistic simulation of

the flood event and thus makes this kind of scenario studies a useful help
for decision support in the field of flood management.

References

1. A. W. Hesselink, G. S. Stelling, J. C. J. Kwadijk and H. Middelkoop, Inundation of a Dutch river polder, sensitivity analysis of a physically based inundation model using historic data, *Water Resources Research* **39**, 9 (2003).
2. G. S. Stelling, H. W. J. Kernkamp and M. M. Laguzzi, Delft Flooding System:
   a powerful tool for inundation assessment based upon a positive flow simulation, in *Hydroinformatics '98*, eds. V. Babovic and L. C. Larsen (Balkema,
   Rotterdam, the Netherlands, 1998).
3. J. F. Dhondia and G. S. Stelling, Application of one-dimensional-two-
   dimensional integrated hydraulic model for flood simulation and damage
   assessment, in *Hydroinformatics 2002: Proceedings of the Fifth International
   Conference on Hydroinformatics. Volume One: Model Development and Data
   Management*, eds. R. A. Falconer, B. Lin, E. L. Harris and C. A. M. E. Wilson,
   (IWA Publishing, London, 2002), pp. 265–276.
4. K. Nippon, River Basin and Watershed Management Program, Water
   Resources Management Plan Formulation and Phase I Project Feasibility
   Study for the Bicol River Basin. Unpublished WorldBank/Nippon Koei report,
   2003.
5. National Economic and Development Authority (NEDA), Results of a
   MIKE11 1D-hydraulic model for the Bicol Basin, the Philippines (Personal
   Communication), 2004.
6. Penning-Rowsell and Tunstall, Risks and resources: defining and managing the floodplain, Chapter 15, *Floodplain Processes* eds. M. G. Anderson,
   D. E. Walling and P. D. Bates (Wiley, New York, 1990).

NUMERICAL ESTIMATION OF PUMP STATIONS EFFECTS UNDER HEAVY RAINFALL IN LOW-LYING AREA USING INUNDATION FLOW ANALYSIS

KENJI KAWAIKE*, HIROTATSU MARUYAMA and MASATO NOGUCHI

Department of Civil Engineering, Nagasaki University
1-14, Bunkyo-machi, Nagasaki, Japan
*kawaike@civil.nagasaki-u.ac.jp

In this study, a comprehensive inundation flow model, which comprises one-dimensional unsteady flow model, two-dimensional inundation flow model and runoff model, is developed and applied to the Isahaya low-lying area, Japan. A numerical simulation using the observed rainfall data of 1999 is conducted, and good agreement between the simulation result and actual inundated area is obtained. Then, as the application example of this model, the effect by extending pump capacity on the inundated area is discussed by estimating the relationship between additional pump capacity and inundated area.

1. Introduction

Recently in Japan, flood disasters due to heavy rainfall occur frequently in highly urbanized area. As the countermeasures against this kind of disasters, not only structural countermeasures but also nonstructural ones are adopted. As an example, "hazard maps" for flood disasters are published from many local governments in Japan, which includes both potential dangerous areas for flood disasters by using numerical simulation and evacuation information such as shelters and evacuation routes. Therefore, an accurate simulation model of inundation flow analysis becomes more necessary as one of the countermeasures against flood disasters.

In this study, a numerical model, which can simulate inundation flow due to heavy rainfall, is proposed and applied to the Isahaya low-lying area, Japan. As the practical application example of this model, the effect of hypothetical pump capacity on inundated area is discussed.

2. Isahaya Low-Lying Area

Isahaya low-lying area is located in the western part of Japan as shown in Fig. 1. It has a problem of rainwater drainage and has been suffering from frequent inundation disasters due to heavy rainfall.

In 1999, heavy rainfall hit the Isahaya City. The temporal change of rainfall intensity observed at Isahaya City is shown in Fig. 2. The maximum rainfall intensity was 95 mm/h, and total rainfall amounted to 347 mm. This heavy rainfall caused an inundation disaster and large part of the low-lying area was inundated.

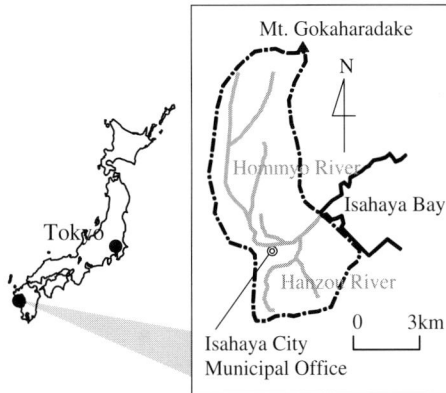

Fig. 1.   Location of the Isahaya city.

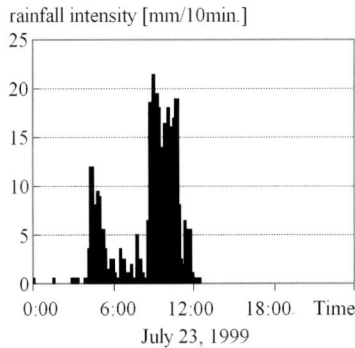

Fig. 2.   Rainfall intensity.

3. Method of Analysis

3.1. *Governing equations*

In this study, the computational area is divided into three parts: river channel, hillside area, and flood-prone area, as shown in Fig. 3.

In the river channel, one-dimensional (1D) unsteady flow analysis using the characteristics method is applied following to Inoue *et al.*[1]:

$$\frac{\partial A}{\partial t} + \frac{\partial Q}{\partial x} = q,$$

$$\frac{1}{g}\frac{\partial u}{\partial t} + \frac{u}{g}\frac{\partial u}{\partial x} + \frac{\partial h}{\partial x} = s_0 - s_f,$$

where A is cross-sectional area of flow, Q the discharge, q the lateral inflow from unit length of the x-direction, u the velocity averaged over cross-section, s_0 the river bed slope, s_f the friction slope, and g is the gravitational acceleration.

In the hillside area, runoff analysis using the kinematic wave method integrating surface and subsurface flow[2] is applied:

$$\frac{\partial h}{\partial t} + \frac{1}{b}\frac{\partial(qb)}{\partial x} = r\cos\theta,$$

$$q = \frac{k\sin\theta}{\gamma}h \qquad\qquad (0 < h < \gamma D),$$

$$q = \frac{\sqrt{\sin\theta}}{n}(h - \gamma D)^m + \frac{k\sin\theta}{\gamma}h \quad (h \geq \gamma D),$$

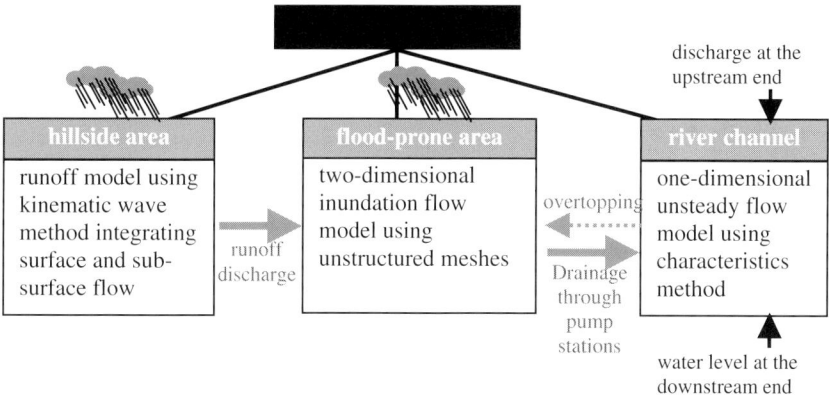

Fig. 3.   Framework of the model.

where h is the substantial water depth on the slope, q the discharge per
unit width on the slope, b the slope width, θ the slope gradient, r the
rainfall intensity, k the hydraulic conductivity, γ the effective porosity, n the
roughness coefficient, D the superficial A-layer depth, and m is the constant
value of 5/3. The obtained runoff discharge at the downstream end of each
slope is given to the flood-prone area as the boundary condition.

In the flood-prone area, two-dimensional (2D) inundation flow analysis,
based on the "unstructured mesh model",[3] is conducted.

$$\frac{\partial h}{\partial t} + \frac{\partial M}{\partial x} + \frac{\partial N}{\partial y} = r,$$

$$\frac{\partial M}{\partial t} + \frac{\partial (uM)}{\partial x} + \frac{\partial (vM)}{\partial y} = -gh\frac{\partial H}{\partial x} - \frac{gn^2 M\sqrt{u^2 + v^2}}{h^{4/3}},$$

$$\frac{\partial N}{\partial t} + \frac{\partial (uN)}{\partial x} + \frac{\partial (vN)}{\partial y} = -gh\frac{\partial H}{\partial y} - \frac{gn^2 N\sqrt{u^2 + v^2}}{h^{4/3}},$$

where h is water depth, M, N are the x- and y-directional discharge per unit
width, respectively, u, v are x- and y-directional flow velocity, respectively,
and H is water stage ($= h + z_b$, z_b is the ground elevation). The rainwater
given to the flood-prone area is drained through pump stations and its
drained water is given to the river channel as the lateral inflow.

3.2. *Application to the Isahaya low-lying area*

The computational reach of the Hommyo River is 5.2 km (from Urayama
to Shiranui), and that of the Hanzou River is 3.1 km (from Umetsu to the
confluence with the Hommyo River). The spatial interval of discretization
Δx is 200 m, and the value of roughness coefficient is 0.030 for the Hommyo
River and 0.045 for the Hanzou River.

The steep slopes used in the hillside area are adjacent areas to the flood-
prone area as shown in Fig. 4. The values of the parameters used here are:
k is 0.002 (m/s), γ is 0.15, n is 0.3 (m$^{-1/3}$s) and D is 0.5 (m).

The computational meshes used in the flood-prone area (shown in Fig. 4)
are divided into four categories: urban areas, channels, streets, and cultivated areas. The values of roughness coefficient of them are 0.067, 0.020,
0.043, and 0.025, respectively.

As the simulation conditions, the flow discharge at Urayama and
Umetsu, and the water level at Shiranui observed at the disaster of 1999
are given to the river channel. The temporal change of rainfall intensity
observed at the disaster of 1999 (see Fig. 2) is uniformly given to the whole
computational area. The computational time step Δt is set to be 0.05 s.

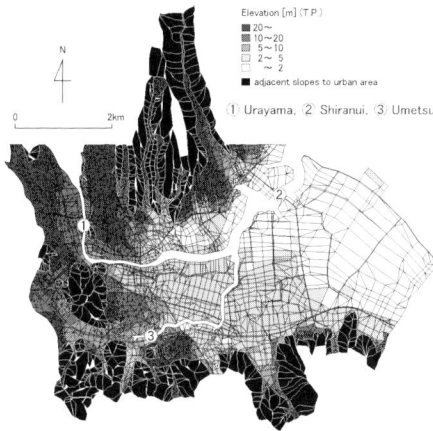

Fig. 4.   Slopes and computational meshes.

4. Results and Discussions

4.1. *Reproduction of the inundated area at the disaster of 1999*

Figure 5 shows the comparison between the actual inundated area and the simulation result (more than 0.1 m water depth). From this figure, the center part of the computational area has a good agreement between the actual inundated area and the simulation result. But the inundated areas of the western and eastern part of the computational area are not well expressed by the simulation. The reasons of these disagreements are that the rainfall spatial distribution is inappropriate and that drainage system by the sewerage is not fully considered in this simulation.

Consequently, using this simulation model, sufficiently good results can be obtained at least around the center part of the computational area.

4.2. *Effects of additional pump capacity on inundated area*

From the above simulation results, three zones of A, B, and C are determined as shown in Fig. 6. Hypothetical construction of new pump stations and extension of existing pump capacity, shown in Fig. 6, are assumed here. The relationship between this additional pump capacity and inundated areas of A, B, and C zones are individually computed and shown in Figs. 7(a)–(c), respectively.

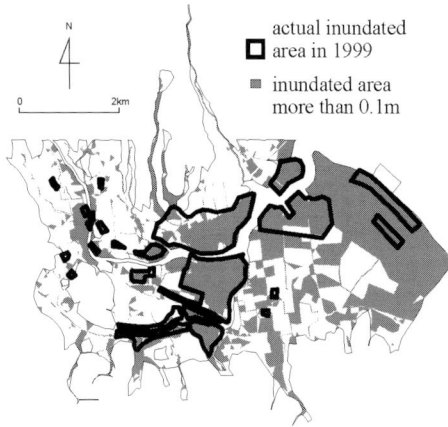

Fig. 5.   Comparison between the actual inundated area and the simulation result.

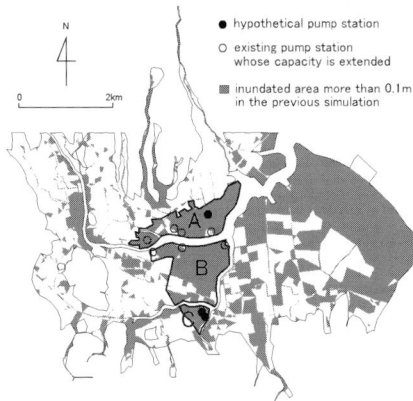

Fig. 6.   Assumed pump stations.

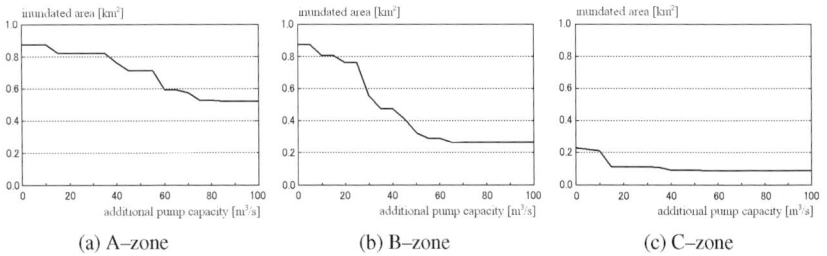

(a) A–zone   (b) B–zone   (c) C–zone

Fig. 7.   Relationship between additional pump capacity and inundated area.

In any zone, additional pump capacity of $100\,\mathrm{m^3/s}$ reduces the inundated area by half. But even additional pump capacity of more than that does not make so significant difference. In this way, useful information to determine the flood prevention planning is available from this simulation results.

5. Conclusions

In this study, a comprehensive simulation method of inundation flow of the Isahaya low-lying area is proposed. Using this model, the dangerous areas for heavy rainfall can be designated, and the effects of newly planned flood-control facilities on the inundated area can be estimated.

References

1. K. Inoue, K. Toda and O. Maeda, Overland inundating flow analysis for Mekong Delta in Vietnam, Hydrosoft 2000, *Hydraulic Engineering Software*, Vol. VIII, (WIT Press, 2000), pp. 123–132.
2. Y. Tachikawa, A. Haraguchi, M. Shiiba and T. Takasao, Development of a distributed rainfall-runoff model based on a tin-based topographic modeling, *Journal of Hydraulic, Coastal and Environmental Engineering*, JSCE **565**, II-39 (1997) 1–10 (in Japanese).
3. K. Kawaike, K. Inoue and K. Toda, Inundation flow modeling in urban area based on the unstructured meshes, Hydrosoft 2000, *Hydraulic Engineering Software*, Vol. VIII (WIT Press, 2000), pp. 457–466.

MODELING OF FLOODPLAIN INUNDATION PROCESS IN LOW-LYING AREAS

PHAM THANH HAI*, TAKAO MASUMOTO and KATSUYUKI SHIMIZU

National Institute for Rural Engineering
2-1-6 Kan-nondai, Tsukuba, Ibaraki, Japan
thanhhai@nkk.affrc.go.jp

A simulation model, which covers floodplains of the Mekong River from Kratie in Cambodia to near the Vietnamese border, was developed by using a finite element method with two-dimensional (2D) shallow-water equations. The model was applied to typical flows of the years 2000 and 2003 (as representative of recent large flood and drought years, respectively) in the river and floodplain systems. Refined, unstructured, triangular FEM meshes were generated for the study area. Main roads, dikes, colmatages, and waterway-opening works, which may have some influence on flow-regimes, were taken into account in the simulation. Moving boundary problem was treated by applying a threshold technique in which nodes having a thin water depth are reset to dry ones in all moving boundary elements every time. Finally, simulated results and observed water-levels and discharges at available gauges were compared to verify the model simulation.

1. Introduction

The existence of reverse flow into Tonle Sap Lake from the Mekong River is a unique feature of the flow in and around the lake and its environs in the Mekong River system. This phenomenon occurs annually; that is, in May/June, the water from the main Mekong River flows down to the lower Mekong and Bassac rivers, and at the same time flows upstream into Tonle Sap Lake. In September/October, when the water level of Tonle Sap Lake is higher than that of the Mekong River, this stored water starts to drain into the lower Mekong and Bassac rivers.[1] The purpose of this paper is to develop a two-dimensional FEM simulation model and to examine this unique feature of flooding processes using full terms of depth-averaged shallow water motion equations. Such models are capable of addressing the geometric complexity usually found in topography as well as at the boundaries of the study area.

2. Governing Equations

Formulation of flow processes can be described by two-dimensional (2D) St. Venant equations for shallow water problems, as follows:

$$\dot{H} + H_{,i}u_i + Hu_{i,i} = 0, \tag{1}$$

$$\dot{u}_i + u_j u_{i,j} - \varepsilon(u_{j,i} + u_{i,j})_{,j} + g(H + z)_{,i}$$

$$+ \frac{gn^2\sqrt{u_k u_k}}{H^{4/3}}u_i - \frac{K\,|W|\,W_i}{H} + fu_i = 0, \tag{2}$$

where $i\,(=1,2)$ or $j\,(=1,2)$ are the subscript for horizontal co-ordinates $(=x,y)$ and t is the time. Differentiations with respect to (x,y) and t are denoted by subscripted comma and superscripted dot, respectively. $H(x,y,t)$ is the water depth; $u_i(x,y,t)$ is the depth averaged velocities; ε is the eddy viscosity coefficient; g is the acceleration of gravity; z is the bed elevation; n is the Manning coefficient of roughness; K is the wind stress coefficient; W is the wind velocity; and f is the Coriolis parameter.

3. Generation of Finite Element Meshes and Solution of the Inundation Processes by FEM

The model is based on 100 m grid-sized DEM topographic data of the study area. And then refined, unstructured-triangular FEM meshes of 62,965 nodes and 124,997 elements were generated. Bed elevations of grids in the DEM data are used to interpolate elevations of FEM nodes in the floodplain domain, while more precise and new updated sound-bathymetry data of the main rivers as bed elevations are utilized to interpolate FEM nodes' elevations in the main rivers' domain. The weighted-residual of the standard Galerkin FEM is applied to the 2D shallow water equations for spatial discretization, and the selective lumping two-step explicit FEM is employed for numerical integration in time. As for initial conditions, the model is run with an assigned zero velocities and water surface elevations with a linear slope based on observed water level data. At land boundaries, slip conditions are imposed along them, so that normal velocities of nodes belonging to land boundaries are set to zero. Observed (or calculated) discharges of 12 tributaries around Tonle Sap Lake are set up at ever-wet nodes which located in water-edge of the Lake in dry season, as inflow-boundaries. Measured water levels as a function of time at Kratie water level gauge are specified for the inflow of upstream conditions, while those at Tan Chau and Chau Doc water level gauges are specified as for the outflow of downstream conditions. Moving boundary problems were treated by applying a threshold technique, where a thin water depth is reset in dry nodes of all moving boundary elements every time (refer Ref. 3 for more detail).

4. Assigning Elevations of Main Roads, Dikes, and Road-Opening Works to FEM Domain

The elevations of those construction works that have a significant influence on flow regimes in the study area were assigned to FEM nodes and elements in the model simulation (see Fig. 1). Construction works were selected based on the "Road Opening Survey" report of Mekong River Commission[2] and other available data, and included NR1 (National-Road 1) from Phnom Penh to the Vietnam border; NR5, NR6, and NR7; a local road on the left bank of the Mekong River, from Khsach Kandal to the junction with NR11 at Ta Kao; dikes on both sides of the Mekong River from Kompong Cham to near Tan Chau; dikes of the Tonle Sap River from Phnom Penh to Prek Dam; dikes on both sides of the Bassac River, from Phnom Penh to near Chau Doc; and main road-opening works located on the selected roads and dikes. Arc-GIS tool was used to convert the polyline shape data of the selected roads, dikes, colmatages, which are a facility of French technology to lead river water into areas behind the levees, and road-opening works to point data. Co-ordinates of these points were used to find the FEM nodes closest to the points, and the elevations of the selected roads, dikes and road-opening works were then assigned to those FEM nodes.

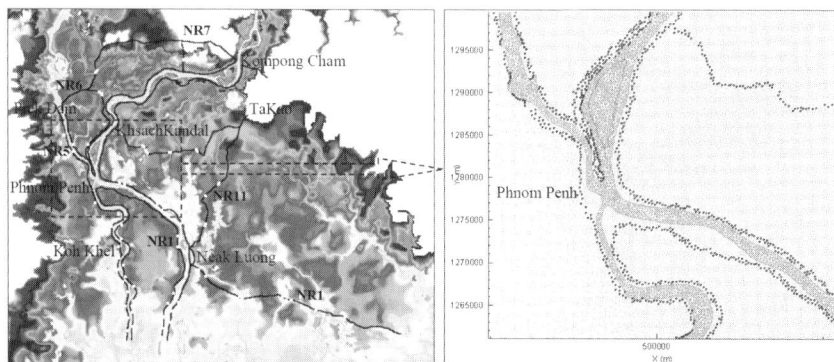

Fig. 1.   Assigning elevations of the selected roads, dikes and road-opening to FEM nodes.

5. Model Simulation Results and Discussion

After applying the methods and parameters mentioned above, the following simulation results were obtained. Figure 2 shows simulated results of the maximum flood extent on September 28, 2000, and that on October 9, 2003. The simulated results reproduced well inundation processes of the year, namely, in drying season water flows only in the main rivers, while during flooding season flood water spills out gradually to adjacent areas, and consequently, flood water expands out both sides' floodplains of rivers.

Based on location of stations and available hydrologic data, we setup five sites to examine calculated water levels, namely Kompong Cham, Kompong Luong, Prek Dam, Koh Khel, and Neak Luong (Fig. 3), and four cross-sections for discharges output, as Kompong Cham, Prek Dam, Koh Khel, and Neak Luong (Fig. 4). Figure 3 shows that the simulation water level at Kampong Cham met well with observed data, while at the others sites they differ to each other; about 0.5–1 m at Kompong Luong, Koh Khel and Neak Luong, and from 1.5–2 m at Prek Dam station. Peak discharge of the model simulation at Kompong Cham ($73,000\,\mathrm{m^3/s}$) is higher than result simulated by MRC ($64,000\,\mathrm{m^3/s}$). Simulated flood hydrograph showed that peaks of flood wave at Kompong Cham is sharp and high while they are flat and lower at Neak Luong, Koh Khel, Prek Dam, that correspond well to real situation of high flood level at upstream and inundation extent at downstream reaches. Result verifications of simulation flow-fields and water levels suggest that, Manning roughness coefficients used in the simulation are small ($n = 0.020$ for main rivers and $n = 0.025$ for floodplains). This makes the simulation flood flows faster than in real state producing high

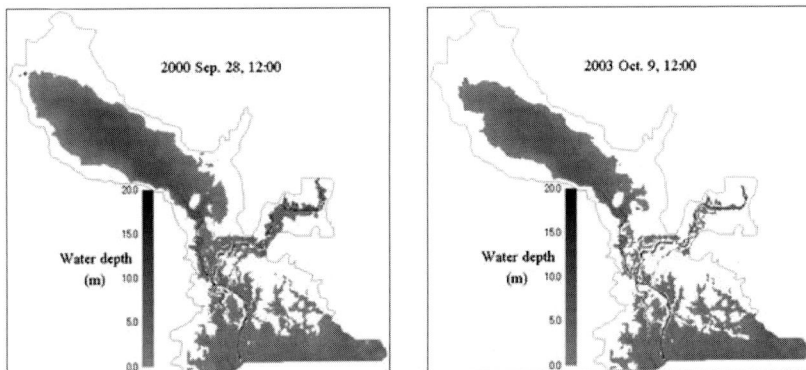

Fig. 2.   Simulated results of the maximum flood extent on September 28, 2000 and that on October 9, 2003.

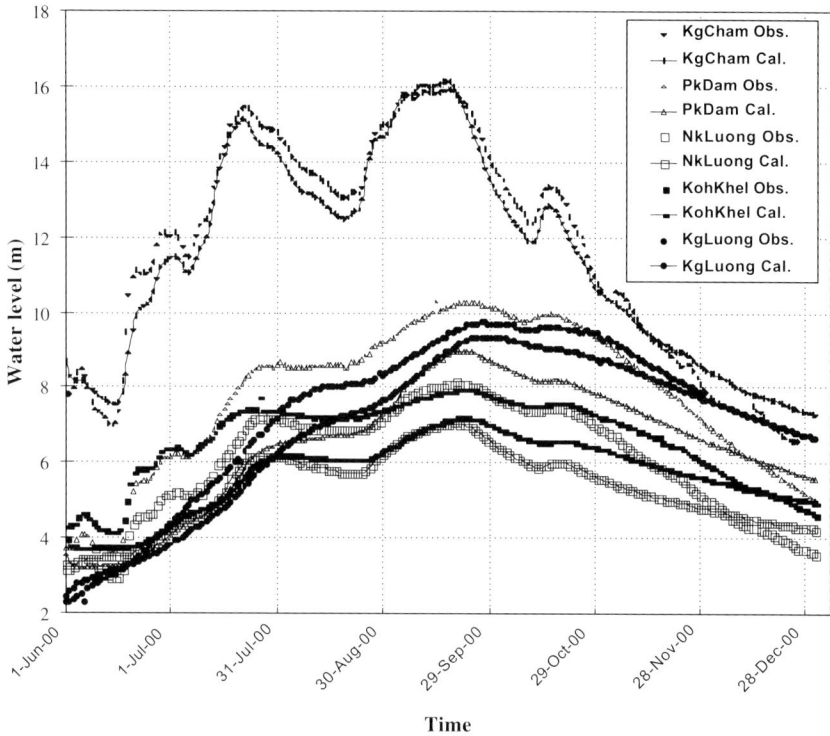

Fig. 3. Simulated water levels compared with observed ones.

discharge, and consequently, simulated water levels are lower than observed data. Therefore, in next simulations we have to try larger Manning roughness coefficients which will retard the flood flows and raise simulation flood water levels. For other reasons for this discrepancy, we are assuming some wrong estimates of mesh size, usage of water level instead of discharge at inflow boundary, discharges of 12 tributaries around Tonle Sap Lake and lumping parameters for FEM calculations.

Results of the model can be used to produce hydrologic data in basin areas without gauging as well as to evaluate the effects of the basin management on floods and agricultural water uses.

Acknowledgments

The authors would like to acknowledge the grant supported from the Revolutionary Research Project "Coexistence of People, Nature and the Earth

P. T. Hai, T. Masumoto and K. Shimizu

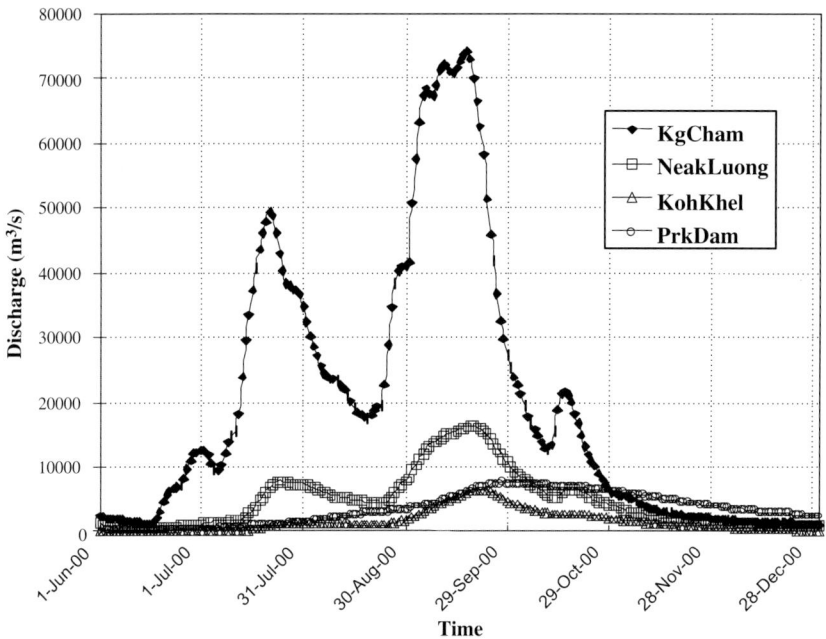

Fig. 4.   Spatial and temporal changes in simulated discharges.

(RR2002)" of the Ministry of Education, Culture, Sport, Science and Technology (MEXT) of Japan. Our thanks also go to the Technical Support Division of Mekong River Commission Secretariat (MRCS) for sharing topographic and hydrologic data, which are used in this study.

References

1. T. Masumoto, *Hydrologic and Environmental Modelling in the Mekong Basin* (Mekong River Commission, 2000), pp. 181–192.
2. Mekong River Commission, *Consolidation of Hydro-Meteorological Data and Multi-Functional Hydrologic Roles of Tonle Sap Lake and its Vicinities* (2003).
3. P. T. Hai, T. Masumoto and K. Shimizu, *Proc. Advances in Integrated Mekong River Manag.* (2004), pp. 339–346.

HEAVY FLOOD DISCHARGE PREDICTION FOR 2004 FUKUI RAINFALL DISASTER IN JAPAN AND PREDICTIONS IN UNGAUGED BASINS

Y. TACHIKAWA*,§, R. TAKUBO†, T. SAYAMA* and K. TAKARA*

*Disaster Prevention Research Institute, Kyoto University, Japan 611-0011

§tachikawa@flood.dpri.kyoto-u.ac.jp

†Department of Urban and Environmental Engineering

Kyoto University, Japan 605-8501

On July 18, 2004, the largest-ever flood since the hydrologic observation began at a catchment occurred at the Asuwa River basin in Fukui Prefecture, Japan. The severe rainfall front in the middle of July brought heavy rainfall with 265 mm in 6 hs. The city area of Fukui was heavily inundated due to dyke breaks along the Asuwa River; and the upper parts of the Asuwa River basin were severely damaged by flood and sedimentation disasters. When designing flood control planning in Japan, to obtain a design flood estimated by using a rainfall-runoff model with a design rainfall having some exceedance probability is fundamental. While the accumulations of hydrologic data to estimate a design flood is insufficient especially in small scale basins. In particular, information of large floods near or above the magnitude of a design level is less available. In this study, we examine how well/bad the largest-ever 2004 flood in Fukui is predicted using a state of the art physically-based distributed rainfall-runoff model; then discuss the source of flood prediction uncertainty and a direction to reduce the uncertainty and enhance the reliability of flood discharge predictions.

1. Introduction

In 2004, severe rainfall fronts and 10 typhoons hit Japan caused heavy rainfall disasters with 232 casualties. These rainfall disasters mainly occurred at tributary catchments with several hundreds square km. River managements of these river catchments are organized by prefectural governments; and in most situations, the river improvements still do not attain a designed safety level. In the future it is still not easy to achieve high river improvements.

In these catchments, to assess the safety level of river improvements at present time is the basis to design a future river development program. Also, to develop a real-time flood runoff prediction system to issue a flood warning is an urgent issue to save lives. To achieve these purposes, hydrologic prediction by a reliable rainfall-runoff model is fundamental. While especially for small scale catchments with several hundreds square km, accumulations of hydrologic data are usually quite insufficient. Generally, the shape of a flood hydrograph at small scale basins is sensitive to space and especially time distributions of rainfall patterns. Therefore, flood runoff predictions at small scale basins request more detailed information of rainfall distribution patterns than at large scale basins with more than several thousand square km. In addition, the flood data with the magnitude of a design flood level or above the level does not exist in most situations. In this sense, rainfall-runoff models are not validated for estimating a flood with a magnitude of design level. To examine the performance of a rainfall-runoff model for the historical 2004 Fukui flood under available hydrologic data is a good test to understand the behaviors and predictability of a rainfall-runoff model and this gives knowledge to improve the prediction of a runoff model.

In this paper, we use a physically based distributed rainfall-runoff model based on topographic representations by grid based DEMs and kinematic flow routing[1-3] and apply the rainfall-runoff model to the upper part of the Asuwa River basin ($351 \, \text{km}^2$); examine the predictability of the model for the 2004 Fukui flood and the source of flood prediction uncertainty; then discuss a direction to reduce the uncertainty and enhance the reliability of flood discharge prediction.

2. Distributed Rainfall-Runoff Model

Figure 1 shows the topographic model of the Asuwa River basin ($351 \, \text{km}^2$) using a DEM processed with the algorithms by Shiiba et al.[1] A slope segment is represented by a rectangle formed by the adjacent two grid points determined to have the steepest gradient. The catchment topography is represented as connections of these slope segments. The spatial resolution of a DEM used here is $50 \, \text{m}$.

According to the flow directions shown in Fig. 1, the slope flow is routed one dimensionally; the slope discharge is given to the river flow routing model; then the river flow is routed to the outlet. In each slope segment, the slope is assumed to be covered by a permeable soil layer composed of a capillary layer and a noncapillary layer. In these layers, slow flow and quick flow are modeled as unsaturated Darcy flow with variable hydraulic conductivity and saturated Darcy flow. If the depth of water exceeds the soil water capacity, overland flow happens. These processes are represented with a kinematic wave model using the continuity equation and a function

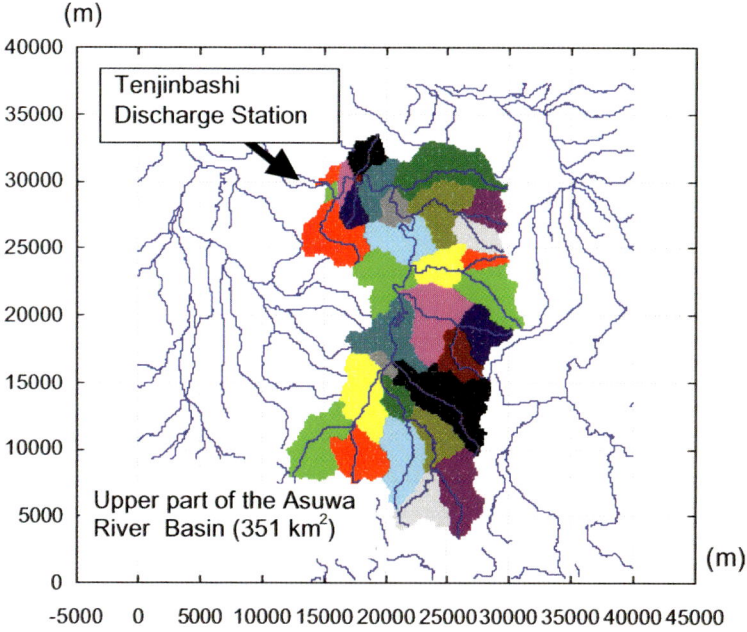

Fig. 1. Watershed model for the upper part of the Asuwa River basin with the outlet at the Tenjinbashi discharge station and its sub-basins which contribute to the river segments of the Asuwa River. The locations are specified using UTM coordinate with m unit.

of the discharge-stage relationship[3] as:

$$q = \begin{cases} v_m d_m (h/d_m)^{\beta}, & 0 \leq h < d_m, \\ v_m d_m + v_a(h - d_m), & d_m \leq h < d_a, \\ v_m d_m + v_a(h - d_m) + \alpha(h - d_a)^m, & d_a \leq h, \end{cases} \tag{1}$$

where q is discharge with unit width; h the flow depth; $v_m = k_m i$, $v_a = k_a i$, $k_m = k_a/\beta$, $\alpha = \sqrt{i}/n$; i the gradient of slope segment; k_m the saturated hydraulic conductivity for the capillary soil layer; k_a the hydraulic conductivity for the noncapillary soil layer; n the surface roughness coefficient; d_m the capacity of water depth for the capillary soil layer; and d_a is the capacity of water depth including capillary and non-capillary soil layer. The model parameters to be determined are $n\,(\mathrm{m}^{-1/3}\mathrm{s})$, $k_a\,(\mathrm{m/s})$, $d_a(\mathrm{m})$, $d_m\,(\mathrm{m})$, and β. For river flow routing, surface flow with rectangular cross section is assumed for kinematic wave approximation.

3. Parameter Estimation and Flood Simulations

The nine floods with more than $400\,\text{m}^3/\text{s}$ peak discharge were selected from the discharge data since 1978. For each flood, spatial distributions of hourly rainfall with 3 km gird resolution were generated from the ground gauged rainfall measurements using the nearest-neighbor method; then the five model parameters above mentioned that define the stage-discharge relationship are determined. The initial water depth at each slope segment is determined from the initial river discharge at the outlet assuming a steady-state condition. Table 1 summarizes the results of parameter identifications. To evaluate appropriateness of the simulated discharges, the peak discharge ratio and the Nash-Sutcliffe efficiency are used. The evaluation results suggest that the identified model parameter sets are classified into three groups: group 1 with the 1993, 1981, and 1982 flood; group 2 with the 1985, 1983, and 1979 flood; and group 3 with the 1989, 1990, and 2004 flood. The parameter sets in group 1 have tendency to overestimate and group 3 to underestimate floods; group 2 shows both tendencies. The estimated peak discharges are widely distributed from 2,500 to $4{,}200\,\text{m}^3/\text{s}$, and it is larger than the estimated peak discharge $2{,}400\,\text{m}^3/\text{s}$ obtained by the Construction Ministry from hydraulic river flow simulations with the high flood stage marks.

4. Discussion

Within a group, the values of model parameters are close, and floods simulated with any parameter sets show good scores of the peak ratio and the Nash efficiency. The clear difference among the groups is that the value of β is larger and the capacity of non-capillary layer d_a–d_m is smaller in group 1 as compared to group 3. The difference of parameter values represents the difference of hydrologic characteristics. In group 1, rainfall stored in soil layer flows quite slow, therefore at the beginning of floods, river discharge is insensitive to rainfall intensity. Then the soil layer is easily saturated and once the water depth at slope segments exceeds the capacity of the capillary soil layer, river discharge rises up suddenly. On the contrary, the parameter sets in group 3 tends to show opposite characteristics that a hydrograph rises up from the beginning of rainfall and its peak discharge is smaller when the same rainfall is given to the runoff model with parameter sets in group 1. Group 2 shows the middle feature of groups 1 and 3.

The question is why the difference is observed in the same catchment. Each group includes various scales of floods and there are no distinguished

Table 1.   Model parameter values fitted to each year flood and the characteristics of each year heavy rainfall and flood discharge.

Properties	Group 1: overestimating peak discharge			Group 2: over/underestimating peak discharge			Group 3: underestimating peak discharge		
Parameters	1993	1981	1982	1985	1983	1979	1989	1990	2004
n (m$^{-1/3}$s)	0.4	0.4	0.4	0.4	0.4	0.4	0.4	0.4	0.4
k_a (m/s)	0.01	0.03	0.01	0.01	0.03	0.01	0.01	0.01	0.01
d_a (m)	0.25	0.4	0.2	0.2	0.6	0.17	0.25	0.325	0.26
d_m (m)	0.18	0.35	0.15	0.1	0.15	0.1	0.18	0.2	0.16
$d_a - d_m$ (m)	0.07	0.05	0.05	0.1	0.45	0.07	0.07	0.125	0.10
β (−)	24	24	12	8	12	4	4	8	4
Peak discharge (m^3/s)	548	1117	676	542	758	622	608	447	2400
Initial discharge (m^3/s)	11	65	18	86	36	37	44	17	25
Six hours rainfall R_{6h} (mm)	60	73	42	43	54	84	67	56	265
Two days rainfall R_{2d} (mm)	116	163	136	116	169	103	174	127	297
Rainfall ratio R_{6h}/R_{2d}	0.52	0.45	0.31	0.37	0.32	0.82	0.39	0.44	0.89
Station number	10	4	7	10	4	6	10	10	12

94   *Y. Tachikawa et al.*

characteristics in the rainfall and discharge data to form three groups. One of clear difference is the number of rainfall observatories. All floods in group 3 were observed with more than 10 rainfall stations. It is inferred that the accuracy of rainfall observations affects the value of the tuned model parameters and makes large prediction uncertainty.

Another possible reason that makes the large prediction uncertainty is the setting of initial condition for prediction simulations. Groups 1 and 2 have a flood observed with 10 rainfall stations. The difference among the floods with more than 10 rainfall observations is the initial discharge. The 1993 flood has the smallest initial discharge; while the 1985 flood has the largest initial discharge in these floods. For the 1985 flood, the recession of river discharge is clearly observed. This implies it was not correct for the 1985 flood to assume the steady state condition at the beginning of simulation. Rainfall is spatially distributed and the distributions are memorized in the spatial distribution of soil moisture, therefore if the initial condition setting is inappropriate the resultant model parameter values are obtained wrongly.

For the 1993 flood, a sudden rising up of river discharge followed dry condition. To simulate the flood with the runoff model used here, the value of parameter β is needed to be set in a large value to keep water in soil layer for lasting small discharge at the outlet. If the low flow observation in 1993 is correct, the improvement of model structure including the refinement of the discharge stage relationship and the initial condition setting are the key to improve the flood runoff prediction for the rainfall-runoff model.

References

1. M. Shiiba, Y. Ichikawa, T. Sakakibara and Y. Tachikawa, *J. Hydraulic Coastal and Environmental Engineering* JSCE **621**/II-47 (1999) 1–9.
2. Y. Ichikawa, M. Murakami, Y. Tachikawa and M. Shiiba, *J. Hydraulic Coastal and Environmental Engineering,* JSCE **691**/II-57 (2001) 43–52.
3. Y. Tachikawa, G. Nagatani and K. Takara, *Annual Journal of Hydraulic Engineering* JSCE **48** (2004) 7–12.

DROUGHT ANALYSIS AND COMPARISON
OF METHODS: A CASE STUDY OF WESTERN UGANDA

A. I. RUGUMAYO* and J. M. MAITEKI[†]

*Department of Civil Engineering Makerere University, Uganda
†National Water and Sewerage Corporation, Uganda

There are several methods of analyzing meteorological drought. These include the drought volumes method and the Standard Precipitation Index. In the former, deficiencies are calculated between the expected and actual rainfall. In the latter, the ratio of the difference between the measured rainfall and the long-term mean to the standard deviation is calculated. In this study, stations that have sufficient years of data were selected. The objectives being to determine the extent of drought in the drought prone areas and derive a relationship between the two methods so that by applying one method, the appropriate values for the other method can be derived, which may be used in future assessments and prediction. The analysis of drought shows that the two regions have similar drought event patterns and that Rakai experiences more severe drought than Masindi. The two methods are related by a linear relationship. This study can be extended to other regions in Uganda.

1. Introduction

Drought is not a new phenomenon in Uganda. Many areas are often hit by severe drought events causing intensive search for water for domestic and agricultural use.[1] Drought events recur continuously and researchers are able to predict their occurrence using the science of meteorology, statistical and probabilistic models. There is no universal definition of drought;[2] thus various perspectives on drought are in existence including meteorological, hydrological, agricultural, socio-economic, and physiological perspectives, amongst many others. The objectives of the study were to study the drought characteristics of two climatic regions using two different methods and compare the methods such that if one method is used results for the other can be obtained. In each case drought is said to occur when the parameter under consideration is lower than the expected amount in a given period. A drought may be simply defined as an abnormal moisture deficiency in relation to some need. This study relates to meteorological drought.

2. Study Area

Masindi district lies between latitudes 10°30'N and 2°30'N, longitudes 31°18'E and 32°18'E while Rakai district lies between latitudes 0° and 1°S, longitudes 31°E and 32°E.

Uganda was delineated into climatological zones by basing on similarity of rainfall characteristics which are shown in Fig. 1.[3] Rakai and Masindi are located such that they spread over many climatological zones.

Masindi district spreads over zones K, L and I while Rakai district spreads over zones CE and A1.

3. Methodology

Rainfall records of all the stations within the study areas were obtained from the Department of Meteorology. The rainfall records were first checked to

Demarcated Rainfall Zones

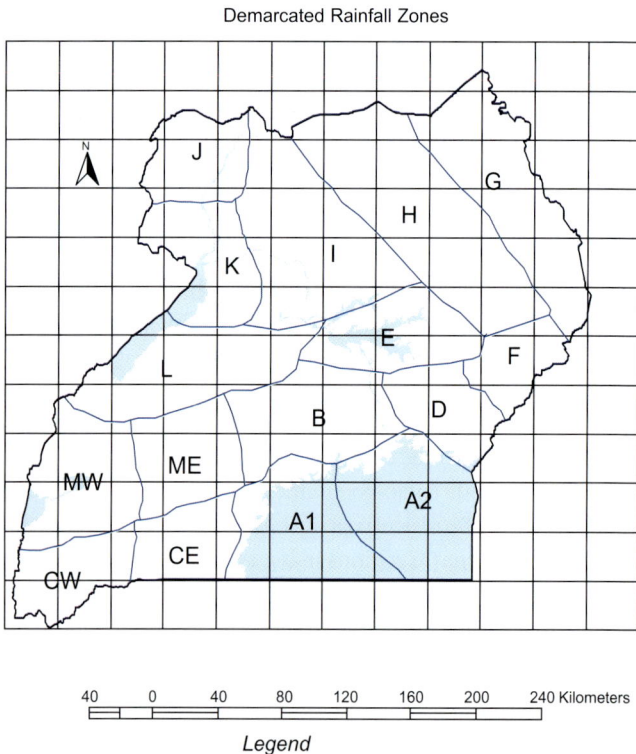

40   0   40   80   120   160   200   240 Kilometers

Legend

Preparedby the GIS SubUnit of the Water Resources Management Department, Entebbe

Fig. 1.   Map showing the climatological regions in Uganda.

see whether their characteristics conform to the respective climatic zones by plotting average monthly rainfall against calendar months; and comparing whether the resulting pattern is similar for all stations in the same climatic zone. The station with the longest record and least missing data, which is the most reliable station, was chosen for the analysis.[4] Masindi Meteorological station was chosen for Masindi district, where as Kalisizo Meteorological station was chosen for Rakai district.

The missing data records were in-filled using the Normal Ratio method and the data were extended using the Markov Generation Techniques.[5] The data were then checked for homogeneity using the Double Mass Curve which revealed consistency in the sets of data. The climate trends of the two regions were established from the rainfall and temperature patterns of the regions. Drought analysis was done using two methods namely Standardized Precipitation Index (SPI) method and the Drought Volumes Method.

3.1. *Standard precipitation index method*[6]

The Standard Precipitation Index (SPI) value is defined as the ratio of the difference between the measured rainfall and the long-term mean to the standard deviation for any month (as shown in Eq. (1)).

$$\text{SPI} = (R_t - M_t)/\sigma_t, \tag{1}$$

where σ_t is the standard deviation for the month, R_t and M_t are the measured monthly and long-term mean monthly rainfall respectively.

Positive SPI values indicate greater than mean precipitation, and negative values indicate less than mean precipitation thus a deficit. The drought magnitudes were calculated as the positive sums of the SPI values that are continuously negative over consecutive months.

3.2. *Drought volumes method*[7]

Drought volumes were calculated by cumulating consecutive rainfall deficits (Eq. (2)). Rainfall deficits were the differences between effective precipitation and the mean monthly rainfall. The drought volumes were then used to calculate the respective drought intensities (Eq. (5)), severity indices (Eq. (7)) and return periods (Eq. (8)).

$$V = \sum_{i=1}^{D} (E_t - M_t)_i, \tag{2}$$

$$E_t = (R_{t-1} - M_{t-1}) * W_t + R_t, \tag{3}$$

$$W_t = 0.1 \left[1 + \frac{12M_t}{MAR} \right], \tag{4}$$

where E_t is the effective precipitation defined by Eq. (3), R_t is the monthly rainfall, M_t the monthly mean rainfall and W_t the weighting factor, which allows for carry-over from one month to the next (Eq. (4)) and MAR is the mean annual rainfall. D is the duration of drought.

The drought intensity is given by

$$Y = \frac{\sum_{i=1}^{D} \left[(E_t - M_t) - (MMD)_t \right]}{\sum_{i=1}^{D} (MMD)_t}, \tag{5}$$

where $(MMD)_t$ is the monthly mean deficit ($t = 1, 2, \ldots, 12$) evaluated from the N years of record and is given by

$$(MMD)_t = \sum_{i=1}^{N} \frac{(M_t - R_t)_i}{N}. \tag{6}$$

The sum $\sum (MMD)_t$ gives the mean annual deficit MAD.

A severity index of drought $= YD$. \hfill (7)

Return period T_r is given by

$$T_r = (n+1)/m, \tag{8}$$

where m is the rank and n is the number of events.

4. Analysis and Discussion

4.1. Standard precipitation index method

Table 1 shows the most severe droughts for both areas. The plots for the drought magnitudes against return periods are shown in Figs. 2 and 3. From the plots, it is clearly seen that there is a linear relationship between the drought magnitudes and the return periods associated with them.

Furthermore, large magnitudes are associated with very large return periods.

4.2. Drought volumes method

The drought volumes calculated using Eq. (2) were used to determine the drought intensity and severity. The drought volumes were divided into a series of class intervals to develop a duration curve shown in Fig. 4 to illustrate the frequency of occurrence of droughts.

Table 1. Most severe drought periods for Masindi and Rakai.

Year	Months	Duration	Magnitude	Return period
Rakai				
1966	May–Oct.	6	7.778	186
1970–1971	May–Jan.	9	7.417	93
1956	Mar–Nov.	9	6.866	62
1980	Jun–Dec.	7	6.503	47
1948-1949	Sept–Mar.	7	6.288	37
Masindi				
1987–1988	Nov–Jul.	9	7.856	159
1986–1987	Jul–Feb.	8	7.013	80
1956	Feb–Nov.	10	6.959	53
1966–1967	Oct–May	8	6.446	40
1963	May–Oct.	6	6.416	32

SPI Drought Plot for Rakai

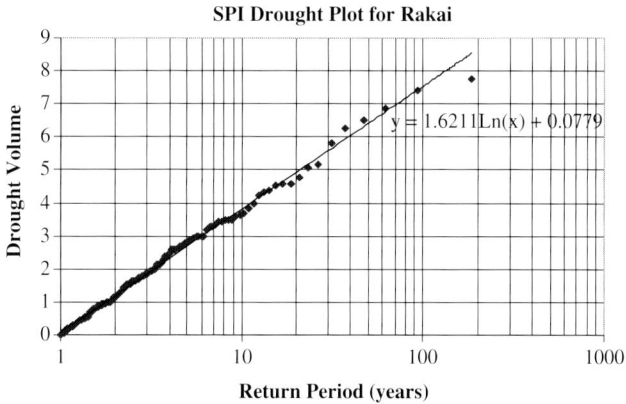

Fig. 2. SPI drought plot for Rakai.

The drought volumes were assumed to be from a population of random events forming a series, $V_1, V_2, V_3, \ldots, V_n$, where n is the number of drought occurrences in the study period in order to apply probability concepts to drought volume analysis. Then a probability distribution was chosen using the Kolmogorov–Smirnov (KS) goodness of fit test. The KS test confirmed that the data best fits to the Log normal distribution. This result is similar to previous drought studies in Uganda.[8–10]

Since the drought volumes best fitted to the Log normal distribution, Eq. (8) was used to estimate the return periods. When log-log plots of drought volumes against return periods where plotted, the relationship in Fig. 5 was obtained.

SPI Drought Plot for Masindi

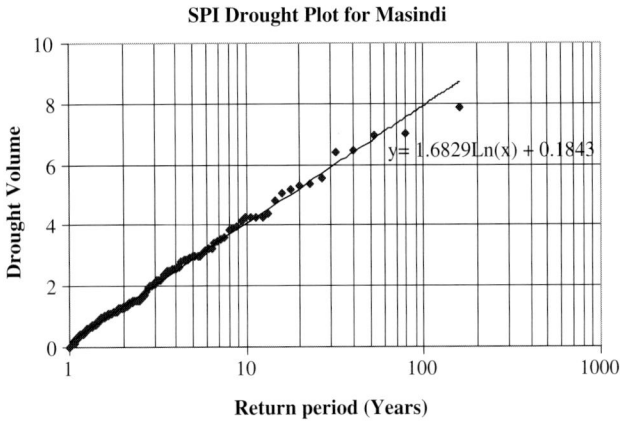

Fig. 3.   SPI drought plot for Masindi.

DURATION CURVE

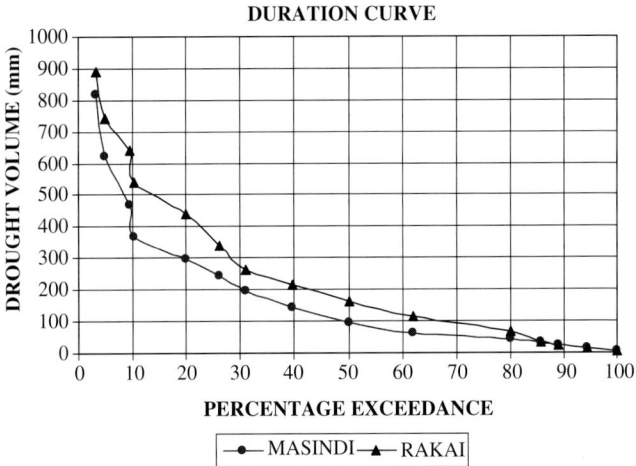

Fig. 4.   Drought duration curve for Masindi and Rakai.

The worst drought in Masindi occurred in the period of January to
June 1983 with a severity of 16.33. The overall deficit in these 6 months
was 578 mm. In Rakai, the worst drought occurred in the period of June
1953 to August 1954 with a severity index of 16.38. The calculated deficit
in these 15 months was 947 mm. It should be noted that the drought with
the longest return period is not necessarily the most severe drought. The
periods with the highest drought volumes experience very severe droughts.

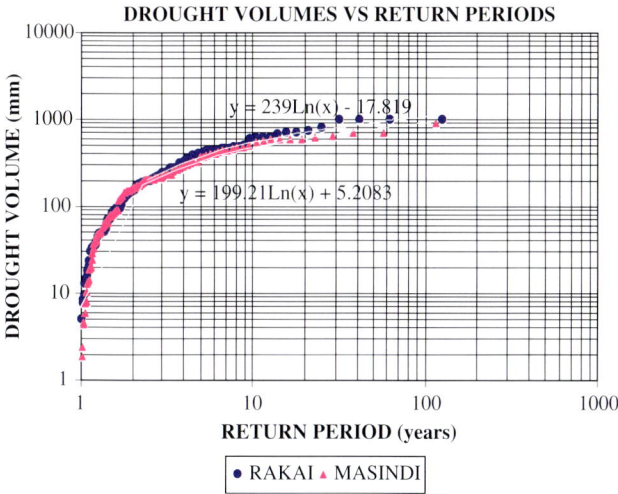

Fig. 5.   Drought volumes against Return period.

Figure 6 shows a positive correlation between the severity index and drought volume for both Masindi and Rakai.

4.3. *Comparison of SPI and drought volumes methods*

Drought volumes and SPI values for various years were cumulated and graphs were plotted to show their respective trends. Figures 7 and 8 clearly show a similarity in the trends for Masindi, whereas Figs. 9 and 10 show the similarity in trends for Rakai.

Figures 11 and 12 show a comparison between the values of SPI and drought volumes in Masindi 1950–1960 and Rakai, respectively. The values can be related using Eqs. (9) and (10).

$$SPI = 0.0044V + 3.485, \quad R = 0.86, \tag{9}$$

$$SPI = 0.0058V + 2.4531, \quad R = 0.79. \tag{10}$$

5. Conclusions

(1) The two regions have similar drought patterns. This is clearly shown by the similarity of the plots for the two regions in Figs. 5 and 6. However, Masindi district experiences relatively more severe droughts than Rakai district. This is shown in Fig. 6, where Masindi district has

A. I. Rugumayo and J. M. Maiteki

DROUGHT VOLUME VS SEVERITY

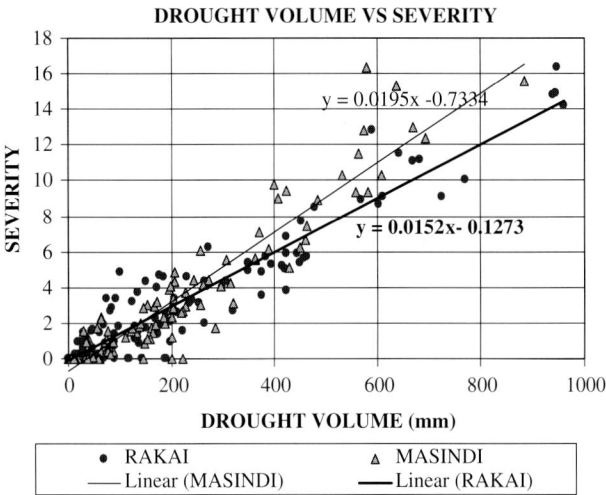

Fig. 6.   Severity against drought volumes.

DROUGHT VOLUMES METHOD

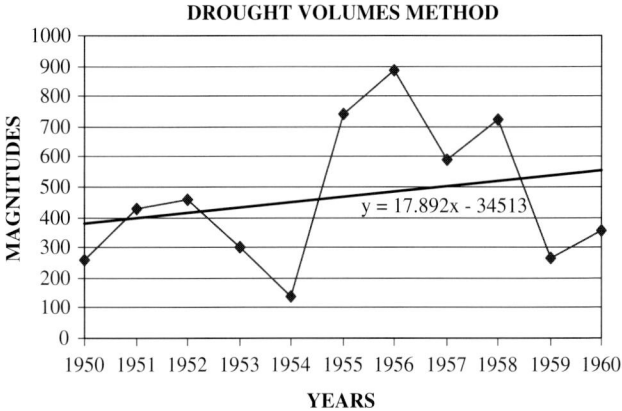

Fig. 7.   Drought volumes against time for Masindi (1950–1960).

a higher severity index than Rakai district for the same drought volume. Furthermore, the droughts for Rakai district have a higher return period than the droughts for Masindi district as shown in Fig. 5.

(2) The Lognormal distribution could be used to study drought in both Masindi and Rakai districts. This distribution satisfied the two

SPI METHOD

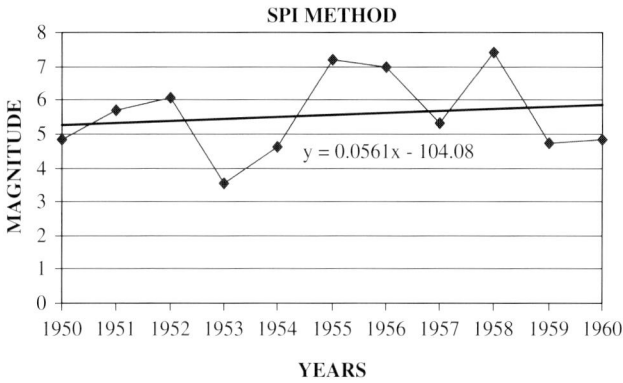

Fig. 8.   SPI method against time for Masindi (1950–1960).

SPI METHOD

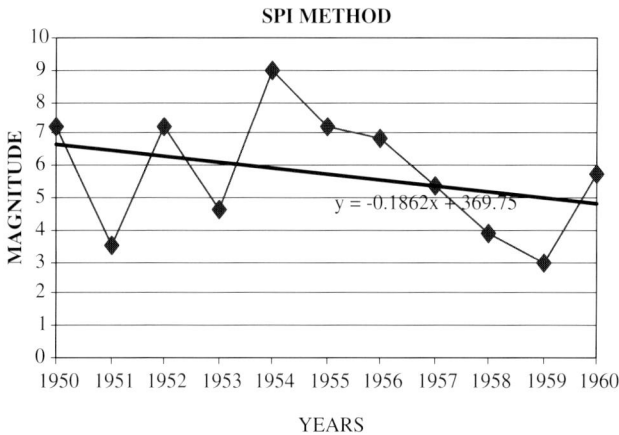

Fig. 9.   SPI method against time for Rakai (1950–1960).

procedures of goodness of fit based on the graphical and KS method carried out on drought volumes.

(3) Comparison of the two methods used in the drought analysis shows that the SPI method is easier than the drought volumes method because it involves less parameters than the latter. However, the droughts volume method gives a more detailed analysis than the SPI method. The two methods can be related by Eqs. (9) and (10). The study can be extended to other regions of Uganda. The relationship between the two methods

A. I. Rugumayo and J. M. Maiteki

DROUGHT VOLUMES METHOD

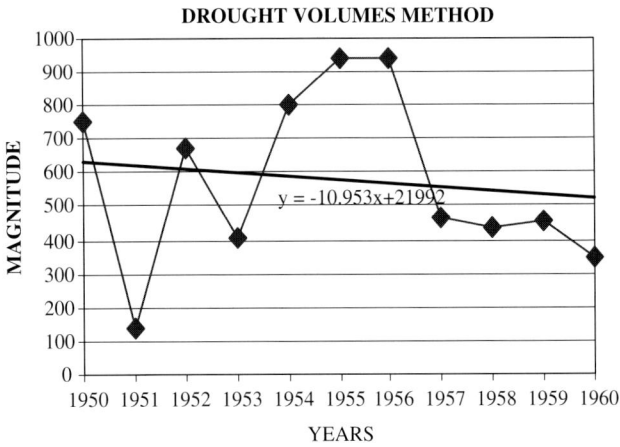

Fig. 10.   Drought volumes against time for Rakai (1950–1960).

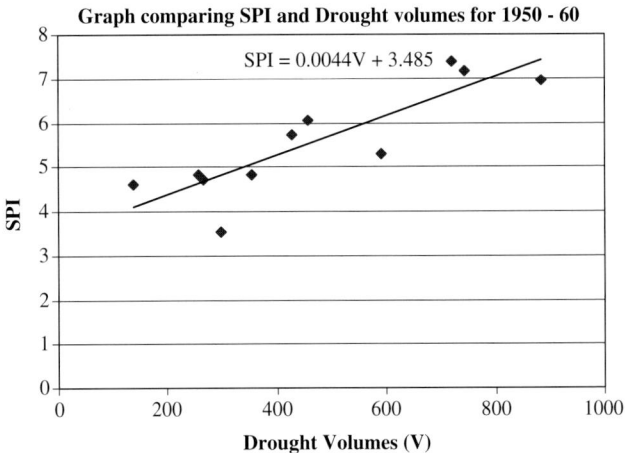

Fig. 11.   Drought volumes and SPI values for Masindi 1950–1960 compared.

should be compared for other regions to examine the extent of the applicability of Eqs. (9) or (10).

Acknowledgments

Special thanks go to the staff of the Department of Meteorology, Ministry of Water, Lands and Environment for their assistance during the course

Graph Comparing SPI and Drought Volumes 1950 - 1960

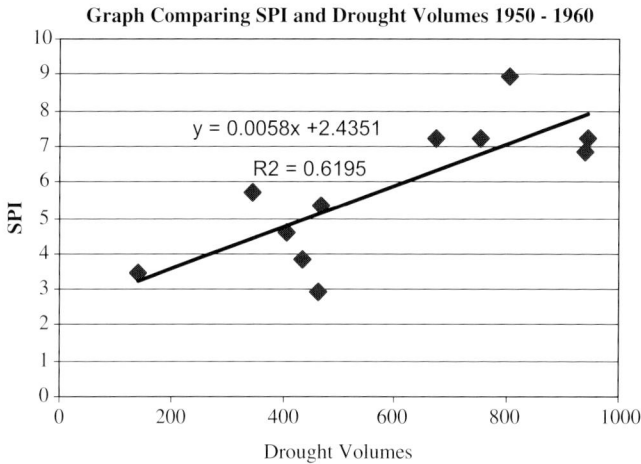

Fig. 12.   Drought volumes against SPI values for Rakai.

of this research. In particular, Mr. A. W. Majugu, Principal Meteorologist and Mr. R. Rukanyangira, Meteorologist at the Department.

Special gratitude is extended to Mr. P. Syayipuma, and Mr. W. Opio, Water Officers Buliisa and Bujenje counties respectively, for their help during the study in Masindi.

References

1. State of Environment Report, National Environment Management Authority 2001, Kampala Uganda, 2000.
2. L. J. Ogallo, Climatological aspects of draught. A conference on draught mitigation, Kampala, Uganda, 1987.
3. C. P. K. Basalirwa, A paper on delineation of Uganda into climatological rainfall zones using the method of principal component analysis. *International Journal of Climatology* **15** (1995) 1161–1177.
4. K. Subramanya, *Engineering Hydrology* (Mcgraw Hill-Tata, Delhi, India, 1995).
5. C. T. Haan, Statistical methods in hydrology, Iowa State University, Iowa USA, 1983.
6. T. McKee *et al.*, Standardized Precipitation Index, Use in Draught Analysis Colorado University, USA, 1993.
7. V. Yeyjevich, Tendencies in Hydrology research and its applications in the 21st Century 1991, Water Resources Management, Vol. 5 (Springer Netherlands), pp. 1–23.

8. A. Rugumayo and Mwebaze, Drought intensity and frequency analysis: A case study of western Uganda, *Journal of Chartered Institution of Water and Enviromental Management* **16** (2002) 111–115.
9. J. Semuwemba, *Analysis and Mitigation of Droughts: A Case Study of Central and North Eastern Uganda*, Unpublished Report Makerere University, Kampala Uganda (2001).
10. A. W. Majugu, A paper on the Elnino update in Uganda. Unpublished, Dept of Meteorology Kampala, Uganda.

MASS TRANSPORT MODELING IN THE UPPER KODAGANAR RIVER BASIN, TAMILNADU, INDIA

N. C. MONDAL* and V. S. SINGH

National Geophysical Research Institute, Hyderabad-500 007, India
ncmngri@yahoo.co.in

Groundwater in Upper Kodaganar River basin, Tamilnadu, southern India, is polluted due to discharge of untreated effluents from 80 functional tanneries. Total dissolved solids (TDS) are observed ranging from 2000 to 30,575 mg/l in open dug wells in the tannery cluster. A mass transport model was constructed to study the pollutant migration. The basin covering an area of 240 km^2 was chosen to construct the groundwater flow model in the weathered part of unconfined aquifer condition. The shallow groundwater potential field computed through flow model was then used as input to mass transport model. MT3D computer code was used to simulate the mass transport model. The mass transport model was calibrated with field observation. Sensitivity analysis was carried out whereby model parameters viz. transmissivity, dispersivity etc. were altered slightly and the effect on calibration statistics is observed. This study clearly indicates that the transmissivity plays a sensitive role than the dispersivity indicating that the migration phenomena are mainly through advection rather than dispersion. The study also indicated that even if the pollutant sources were reduced to fifty percent of the present level, the TDS concentration level in the groundwater, even after two decades, would not be reduced below half fold of TDS of present level.

1. Introduction

The study area, a granitic rock formation in Tamilnadu, southern India, possesses poor groundwater potential and serious contamination of both surface water and groundwater has been reported in this area as a result of uncontrolled discharge of untreated effluents by 80 tanning industries for the last three decades.[1-4] The health of the rural farming community and people working in the tanning industries have been seriously affected and they are suffering from occupational diseases such as asthma, chromium ulcers and skin diseases.[5] About 100-km^2 area of fertile land has lost its fertility. TDS concentration in groundwater at some pockets varies from 17,024 to 30,575 mg/l.[6] As the discharge of effluents is continuing, a prognosis of further pollutant migration is carried out using a mathematical model. A numerical model of the area was developed using the finite difference technique coupled with method of characteristics and it also used to predict TDS migration for the next 20 years. Sensitivity analysis was carried out to identify the parameters, which are influencing the contaminant migration. Sensitivity analysis shows that advection and not dispersion is the predominant mode of solute migration. There are a large number of reports and papers available to describe the solute transport models to study the contaminant migration in the industrial belts, coastal aquifer etc.[7-15] The computer software Visual MODFLOW 3.0 is used for the present study.

2. Background of the Area

The area, an area of 240 km², is a hard rock, drought prone region, which is situated in Tamilnadu, southern India (Fig. 1). It is characterized by undulating topography with hills located in the southern parts, sloping toward north and northwest.[16] The elevation varies from 360 m above mean sea level (amsl) in southern portions to 240 m in the northern part. No perennial streams exit in the area, except for short distance streams encompassing

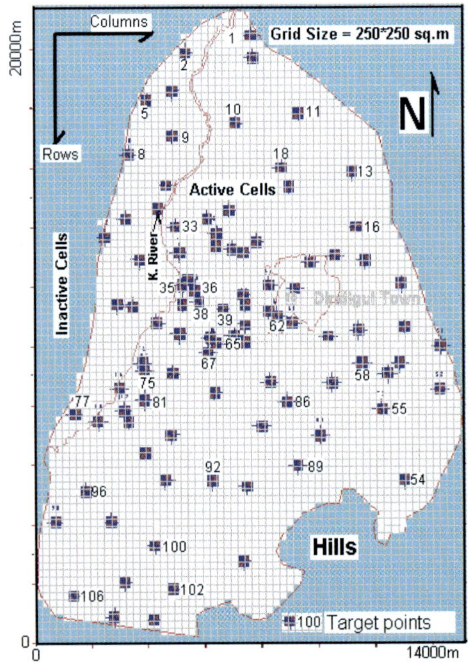

Fig. 1.   Grid map for modeling.

second and third-order drainage.[16,17] The average annual rainfall is about 915.5 mm from a period of 1971–2001.[18]

Geologically, the area is occupied with Archaean granites and gneisses, intruded by dykes. Black cotton soil and red sandy soil (thickness: 0.52–5.35 m) predominate in the area but thickness of weather varies from 3.1 to 26.6 m.[16] Charnokite rock is found in the extreme southern and southeastern part of the Sirumalai hill. Groundwater is extracted through dug well, dug-cum-bore wells and bore wells for different purposes.[19] The general trend of groundwater motion under shallow aquifer is in north and northwest directions.[20]

3. Approach for Numerical Modeling

As geometry and boundary conditions in the aquifer are generally so complex, because aquifer is in hard area.[21] Here the main stages are followed for mass transport modeling. They are: (1) solving the groundwater flow equation using finite-difference method, (2) estimation of fluid velocities at each node, and (3) solving the mass-transport equation using finite-difference technique and method of characteristics using the flow velocities.[22]

Grid design: The area is divided into a series of grid blocks or cells[23] (size of $250 \times 250\,\text{m}^2$, total grids = 3,342, area = $209\,\text{km}^2$) and the groundwater heads will be computed at the center of each grid block (Fig. 1). The layer is unconfined condition and corresponds to a layer type 1 in MODFLOW. Hydraulic conductivity values as well as specific yield values were assigned from the field data. The actual values of the ground surface elevation and bottom elevation of the bedrock were entered at the model.

No-flow boundary: It has been set in the southern part of the basin. Head-dependent boundaries: The northern boundary of the area will be simulated through Generalized Head Boundary (GHB). Other important boundaries: (1) the weathered part of aquifer will be considered as a porous one, (2) areal recharge and pumpages will be assigned at random, and (3) wherever dykes and exposures are present, transmissivity values will be adjusted and assigned as per its direction and length.

TDS in the surface effluents was more than 30,000 mg/l during the period September 1988 to February 2002. The quantity of effluents seeping to the groundwater system was assumed to be 30% of the surface effluents. It was also assumed that on a conservative basis the solvent reaching the water table has a solute concentration, which is 30% of that present at the surface. The remaining 70% of the solutes may get absorbed in the unsaturated zones or are carried away by the runoff. An effective porosity of 0.2, longitudinal dispersivity of 30 m and transverse dispersivity of 10 m were uniformly assumed for the entire area. The mass transport model was calibrated in two stages: steady state and transient state. This transient state model was used for prognostic model. It was also assumed that TDS do not influence the density and viscosity values, which may affect the groundwater flow and pollutant migration. As various other parameters (collected in the field) were assigned to the corresponding nodes.

4. Steady State Calibration

TDS concentration "C" was calculated at all node points for September 1988, a date up to which the system was assumed to be in a steady-state condition. There was a mismatch between observed and computed values of "C". Therefore, efforts were made to obtain a reasonably better match by modifying the magnitude and distribution of the background concentration and pollutant load. However, the situation could not be improved much. This may be due to a variety of factors; the most important which are the lacunae and inaccuracies in the database. To get the real representation of the aquifer system, field data (January 2001) was considered for other steady-state condition and it also run to visualize the mass transport model. The computed versus observed C was given good result.

5. Transient Calibration

As the steady-state model could not reproduce the observed data at all the points, a time variant simulation was carried out. This was done for the period January 1988 to July 1995 based on available of PWD data.[18] The pollution load was reaching the groundwater system at various clusters during this period. The computed "C" for five PWD-wells is higher than the observed values. These values, however, could not be rectified, as there was no basis for modifying either model parameters or the pollutant load in the absence of any actual field estimates. It should be mentioned here that the present model is only to illustrate the feasibility of applying modelling techniques to study this problem and to use it for prediction of system behavior for some future scenarios.

6. Sensitivity Analysis

The impact of varying conductivity, dispersivity, and C/W (TDS pollution load at the source) was studied. The variation is caused in the TDS

Table 1.   Variation TDS concentration for a few target points by varying K, A_L and C'.

Targets	K0C (mg/l)	K1C (mg/l)	K2C (mg/l)	A1C (mg/l)	A2C (mg/l)	C1$'$C (mg/l)	C2$'$C (mg/l)
1	1999.9	1999.9	1999.9	2000.0	2000.0	1998.9	2000.9
16	1928.0	1931.7	1938.5	1925.4	1925.3	1928.3	1928.5
33	9915.5	9909.2	9929.6	9913.3	9910.9	9915.4	9915.5
35	5933.2	5778.8	5864.1	5972.2	5970.8	5933.0	5933.4
36	8316.5	8264.1	8292.5	8330.0	8329.6	8316.0	8316.6
38	8748.4	8629.7	8688.5	8781.8	8782.2	8748.0	8748.9
39	16737.0	16773.0	16758.0	16726.0	16726.0	16736.9	16737.1
58	3004.3	3003.2	3004.6	3004.6	3004.7	3004.0	3004.5
62	3877.8	3889.9	3890.5	3872.7	3872.9	3877.6	3877.8
64	6334.3	6191.1	6198.9	6392.7	6392.5	6336.3	6334.3
75	2308.0	2266.9	2275.5	2321.1	2322.2	2308.6	2308.0
77	3000.0	3000.0	3000.0	3000.0	3000.0	3000.0	3000.0
81	2895.0	2913.8	2903.9	2890.1	2889.9	2895.0	2895.0

($K0$ is conductivity for calibrated model in m/d; $A_L = 30$ m (longitudinal dispersivity); $C' = 9{,}000$ mg/l (concentration); K0C is TDS concentration for $K0$, A_L and C'; K1C is TDS concentration for $K1 = (80\%$ of $K0)$, A_L and C'; K2C is TDS concentration for $K1 = (120\%$ of $K0)$, A_L and C'; A1C is TDS concentration for $A_L = 50$ m, $K0$, C'; A2C is TDS concentration for $A_L = 100$ m, $K0$, C'; C1$'$C is TDS concentration when $C1' = 7{,}200$ mg/l, $K0$, A_L; and C2$'$C is TDS concentration when $C1' = 10{,}800$ mg/l, $K0$, A_L; targets were shown in Fig. 1).

concentration "C" at some selected node points as result of some variations in these parameters is shown in Table 1.

(1) Conductivity: This parameter was changed by 20% (upward and downward) of the value assigned in the model at each node the change in the conductivity affects the groundwater velocity causing redistribution of solute concentration. In general, the higher the conductivity, the faster is the movement of the solute. Therefore, the concentration is reduced near the sources and increased and vice versa (see columns 3 and 4 of Table 1).

(2) Dispersivity: The longitudinal dispersivity was increased to 50 and 100 m (from 30 m). The transverse dispersivity was taken as one-third of the longitudinal dispersivity. No significant changes in the TDS concentration were noticed due to increase in the dispersivity (see columns 5 and 6 of Table 1). This shows that advection and not dispersion is the predominant mode of solute migration in the tannery belt.

(3) TDS pollution load at sources points ($C'W$): The effect of varying this parameter by 20% (upward and downward) at 32 source points (nodes taken at the major tannery clusters) was examined and it was found that TDS concentration "C" rises with an increase in the pollution load $C'W$ and vice versa (see columns 7 and 8 of Table 1).

7. Prognostic Model

The following three scenarios were considered for predicting the extent of pollution in this area at the end of a 20 years period.

(1) The TDS load remains invariant during the entire period of prediction.
(2) The TDS load is increased to two folds of the present level (January 2001) during the entire period of prediction.
(3) The TDS load is reduced to half of the present level. The TDS load is a result of both the effluents discharged from the tanneries and the leaching of the previous adsorbed solutes in the unsaturated zone. Thus effectively the over all discharge from the tanneries is assumed to reduce about 50% of the present level.

The predicted TDS concentration level (Scenario-I) for the year 2020 has shown (Fig. 2) that the TDS concentration "C" progressively increases in the area due to continuous addition of solids to the groundwater system.

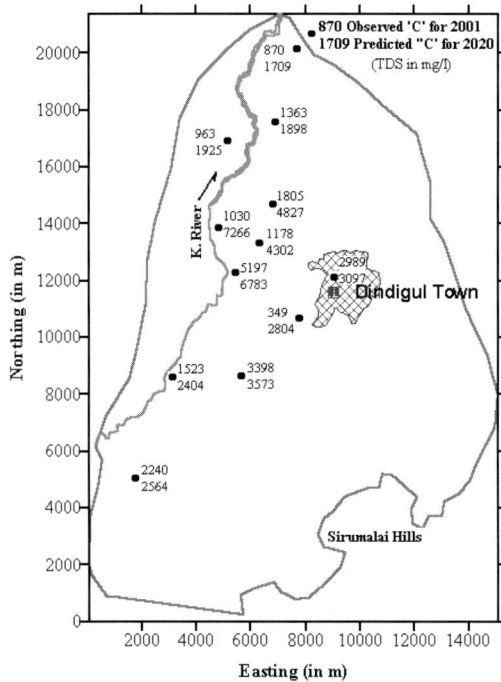

Fig. 2.   Predicted TDS concentration (mg/l, January 2020).

The area, which TDS content in groundwater system may be more than 4,000 mg/l is likely to be doubled within the next two decades from the present size in between river and town towards north and west of Dindigul town.

The Scenario-II (Fig. 3) has shown that that at the end of a 10-year period (2010) TDS concentration "C" will be same of Scenario-I, but may still be quite high at some locations. Scenario-III (Fig. 4) can be seen that at the end of a 20-year period (2020) TDS concentration "C" will be reduced but may still be quite high at some places. At the centre of the tannery cluster TDS concentration is reduced but in the northern side it is increasing order due to movement of the pollutant in advection nature. Prognosis using the model confirms that the polluted area as well as the concentration of pollutants in the groundwater will continue to increase in future. The study also indicated that even if the pollutant sources were reduced to 50% of the present level, the TDS concentration level in the groundwater, even after two decades, would not be reduced below 50% TDS of 2001.

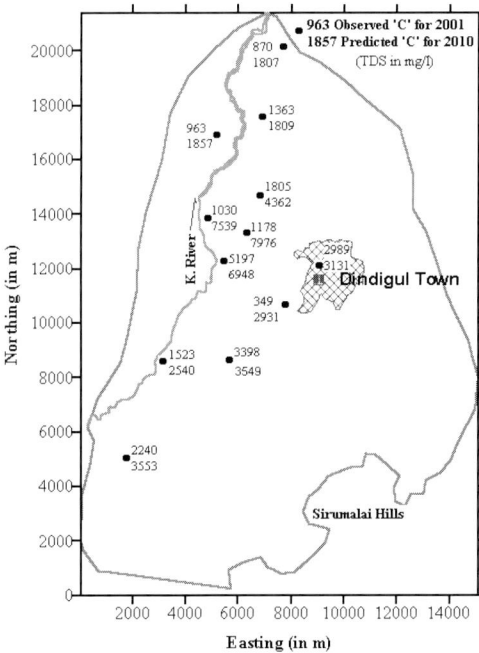

Fig. 3.   Predicted TDS concentration (mg/l, January 2010).

N. C. Mondal and V. S. Singh

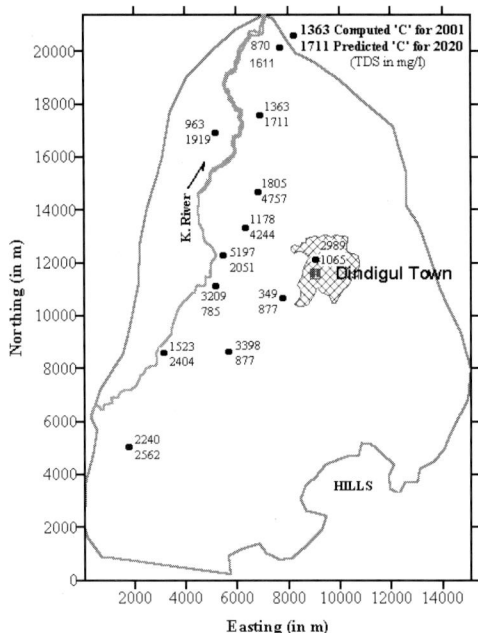

Fig. 4. Predicted TDS concentration (mg/l, January 2020).

8. Conclusions

(1) TDS concentration has been computed through MT3D mass transport model starting with a background concentration 1,000 mg/l. Even through TDS has selected for simulation of contaminant migration, the migration of any species will follow a similar pattern as mass transport is primarily driven by advection.

(2) From transient condition, it is inferred that TDS concentration is steeply increased in and around the tannery cluster. The impact of varying TDS in the tannery belt is based on the advection than dispersive mechanism.

(3) Modeling of pollutant migration in the aquifer is shown that if tannery effluents continue to be discharged at the present level, both as regards the volume and TDS concentration, groundwater pollution will continue to increase.

(4) It is noted that even if tannery effluents are reduced to 50% of the present level, even after 20 years, the TDS concentration in groundwater will not be reduced to half fold of the original level (2001).

However, an exact quantification of the affected area like Dindigul town and concentration of pollutants in groundwater is possible only if one could make a valid model based on a more representative and accurate database.

Acknowledgments

Authors are thankful to Dr. V.P. Dimri, Director, NGRI, Hyderabad, India for permitting to publish this paper. Dr. N.C. Mondal is also thankful to CSIR, New Delhi, India, for their financial support to present this paper in second AOGS Annual Meeting, Singapore.

References

1. N. C. Mondal and V. S. Singh, *Curr. Sci.* **89**, 9 (2005) 1600–1606.
2. N. C. Mondal, V. K. Saxena and V. S. Singh, *Curr. Sci.* **88**, 12 (2005) 1988–1994.
3. N. C. Mondal and V. S. Singh, *Proceedings of International Conference*, WE-2003, Bhopal, *Water & Environment, Ground Water Pollution*, eds. V. P. Singh and R. N. Yadava (Allied Publishers Pvt. Ltd., New Delhi, 2003), pp. 262–277.
4. N. C. Mondal and V. S. Singh, *Proceeding of the Second Asia Pacific Association of Hydrology and Water Resources Conference*, Singapore, Vol.-II (2004), pp. 436–444.
5. J. Paul Basker, *Dossier* (2000) 208–210.
6. N. C. Mondal, V. K. Saxena and V. S. Singh, *Environ. Geol.* **48**, 2 (2005) 149–157.
7. L. F. Konikow and J. D. Bredehoeft, *Water Resources Res.* **10** (1974) 546–562.
8. S. G. Robson, *Water Resources Investigations*, USGS (73-46) (1974) 66.
9. J. F. Konikow, *US Geological Survey Water Supply*, Paper 2044 (US Government Printing Office, Washington, DC, 1976), 1–43.
10. J. F. Konikow and J. D. Bredehoeft, *Techniques of Water-Resources Investigations of the USGS*, Chapter C2, Book 7 (1978) 90.
11. C. P. Gupta, M. Thangarajan, V. V. S. G. Rao, Y. M. Ramachandra and M. R. K. Sarma, NGRI Technical Report No. 94-GW-168, 45, 1994.
12. M. Thangarajan, *Environ. Geol.* **38**, 3 (1999) 209–222.
13. V. V. S. G. Rao and S. K. Gupta, *Environ. Geol.* **39**, 8 (2000) 893–900.
14. A. G. Bobba, *Hydrological Science* **47**, S (2002) S67–S80.
15. P. K. Majumdar, N. C. Ghosh and B. Chakravorty, *Hydrological Science Journal* **47**, S (2002) S55–S66.
16. N. C. Mondal and V. S. Singh, *Proc. Int. Conf. (WE-2003, Bhopal) on Water and Environment, Ground Water Pollution*, eds. V. P. Singh and R. N. Yadava (Allied Publishers Pvt. Ltd., New Delhi, 2003), pp. 188–198.
17. N. C. Mondal and V. S. Singh, *Curr. Sci.* **87**, 5 (2004) 658–662.

116   *N. C. Mondal and V. S. Singh*

18. Public Works Department (PWD), Government of India, Chennai, Report, 102, 2002.
19. V. S. Singh, N. C. Mondal, R. Barker, M. Thangarajan, T. V. Rao and K. Subrahmanyam, Technical Report No. 2003-GW-269, 104, 2003.
20. N. C. Mondal, *Ph.D. Thesis in Geophysics*, Osmania University, Hyderabad, India, 249 (2005).
21. K. R. Rushton and S. C. Redshaw, (Wiley, Chichester, UK, 1979), p. 332.
22. I. Javandel, C. Doughty and C. F. Tsang, *American Geophysical Union Water Resources Monogram* **10** (1984) 228.
23. M. P. Anderson and W. W. Woessner, *Applied Groundwater Modeling; Simulation of Flow and Advective Transport* (Academic Press, London, New York, 1992), p. 381.

MODELING SEDIMENT DISCHARGE WITH ARTIFICIAL NEURAL NETWORK: AN EXAMPLE OF THE LONGCHUANG RIVER IN THE UPPER YANGTZE

YUN-MEI ZHU[*,†], X. X. LU[*], YUE ZHOU[†] and YOUAN GUO[§]

*Department of Geography, National University of Singapore, 119260, Singapore
†Department of Environmental Science, Kunming University of Science and Technology, China
§Yunnan Hydrological and Water Resources Bureau, Yunnan, China

Artificial neural network (ANN) was used to model the monthly sediment discharge in the Longchuang River in the Upper Yangtze, China. The variables including the averaged rainfall and temperature, rainfall intensity and water discharge were used. The results suggest that ANN is capable of modeling the monthly sediment discharge with fairly good accuracy when proper variables and their lag effect on sediment discharge are used as inputs. Compared to the multiple regression models, ANN produced a better fit to the observed sediment discharge and provided more reasonable results for the extreme points.

1. Introduction

Artificial Neural Network (ANN) is based on the concepts derived from the researches on the nature of the human brain.[1] Its distinct advantages make it a competitive tool in hydrological modeling.[2,3] The current application of ANN in hydrology mainly focuses on the river flow modeling and prediction,[4,5] but much less on sediment discharge. The researches conducted by Abrahart and White,[6] Jain,[7] Tayfur,[8] and Kisi[9] may be deemed as pathfinder experiments in this area. They have demonstrated the capability of ANN in sediment concentration or discharge modeling. However, these researches usually took water and sediment discharge at previous time steps as inputs, which may increase the accuracy of the simulation but failed to explain the physical relations between the sediment and its control variables. This research attempted to relate the sediment discharge with the original driving forces such as rainfall, temperature and rainfall intensity, aiming to establish an ANN model that can be used to explore the relationships between the causal variables and the sediment responses. The advantage of the ANN over the multiple linear regression (MLR) models was also evaluated by comparing their performances.

2. Study Area

The study area, the Longchuang River, a tributary of the Upper Yangtze, is located between $24°45'N$–$26°15'N$ and $100°56'E$–$102°02'E$, southwest China. The length of the main channel is 231.2 km. The gauging station, Huangguayuan, has a drainage area of 5,560 km^2 (Fig. 1). The climatic data were collected from six weather stations (Fig. 1). The climatic and hydrologic data in wet season (May–December) from 1963 to 2001 were used for the modeling. Seven variables were selected as the causal variables, including two variables representing the average climate status (monthly average rainfall (R) and temperature (T)), four variables representing the influence from storm event (the numbers of 25 mm- and 50 mm-or-more rainfall days in each month (N_{25}, N_{50}) and their cumulative rainfall (R_{25}, R_{50})) and water discharge (W).

3. Methodology

The original data were processed through three steps for the ANN modeling. First, the climatic data from six weather stations were converted into

Fig. 1.   Location of the Longchuang River and the Thiessen polygons of weather stations.

the mean areal values with the Thiessen method[10] (Fig. 1). Second, all the data were standardized to the interval from 0.1 to 0.9.[11] Third, the data were partitioned into calibrations set (160 records) and validation set (120 records) according to the modified differential split-sample method proposed by Tokar and Johnson.[12]

Multilayer perceptron (MLP) with one-hidden-layer and sigmoid function as the transformation function was chosen as the network. Detailed information about MLPs may be found in the literatures (e.g. Ref. 1). Three groups of neural networks, each consisting of four networks, were constructed. Each group had different types of causal variables and the networks within the same group had different time lags (Table 1). The performances of the networks were evaluated with root mean square error (RMSE) and the coefficient of multiple determination (R^2).[11] The best performing network of each group was identified and their performances were compared to the MLR models which had the same input combinations.

4. Results and Discussion

Twelve ANNs were calibrated, validated, and evaluated. The RMSE and R^2 in the calibration and validation periods of these ANNs are listed in Table 2. ANN_4, ANN_6, and ANN_9 were the best performing networks in their corresponding groups, respectively. Three regression models, MLR_A, MLR_B, and MLR_C, which have the same inputs as ANN_4, ANN_6, and

Table 1. Input combinations of the ANNs.

ANNs		Inputs
Group I	1	$(T, R)t$
	2	$(T, R)t, (T, R)(t-1)$
	3	$(T, R)t, (T, R)(t-1), (T, R)(t-2)$
	4	$(T, R)t, (T, R)(t-1), (T, R)(t-2), (T, R)(t-3)$
Group II	5	$(T, R, R_{25}, N_{25}, R_{50}, N_{50})t$
	6	$(T, R, R_{25}, N_{25}, R_{50}, N_{50})t, (T, R)(t-1)$
	7	$(T, R, R_{25}, N_{25}, R_{50}, N_{50})t, (T, R)(t-1), (T, R)(t-2)$
	8	$(T, R, R_{25}, N_{25}, R_{50}, N_{50})t, (T, R)(t-1), (T, R)(t-2), (T, R)(t-3)$
Group III	9	$(T, R, W)t$
	10	$(T, R, W)t, (T, R, W)(t-1)$
	11	$(T, R, W)t, (T, R, W)(t-1), (T, R, W)(t-2)$
	12	$(T, R, W)t, (T, R, W)(t-1), (T, R, W)(t-2), (T, R, W)(t-3)$

Table 2.   Performance of ANNs and MLR models for Longchuang River.

	Model		Calibration RMSE*	R^2	Validation RMSE*	R^2
	Group I	1	252.07	0.6200	215.42	0.6624
		2	229.47	0.6856	182.59	0.7575
		3	226.26	0.6943	186.83	0.7461
		4	212.96	0.7292	183.01	0.7564
ANN	Group II	5	244.93	0.6305	202.31	0.7023
		6	206.15	0.7382	169.31	0.7915
		7	213.61	0.7189	170.60	0.7883
		8	206.75	0.7367	171.86	0.7852
	Group III	9	190.70	0.7829	133.77	0.8696
		10	192.20	0.7794	134.10	0.8692
		11	190.88	0.7824	145.54	0.8459
		12	188.18	0.7886	139.23	0.8590
MLR		A	233.07	0.6760	234.19	0.6011
		B	225.89	0.6950	226.74	0.6260
		C	200.58	0.7600	186.92	0.7458

*Unit (kg/s).

ANN_9, respectively, were calibrated and validated. Their RMSE and R^2 are given in Table 2.

The selection of the input variables plays an important role in the accuracy of the network. The networks in Group I show that a successful simulation of sediment discharge cannot be made by using averaged rainfall and temperature only as inputs (Table 2). When the variables representing the storm event were added as inputs in Group II, the performance of the network increased significantly. Adding water discharge as input in Group III can further increase the performance of the models. In addition, adding input variables at previous time steps to the network could improve the prediction. However, the degree to which the information from previous months should be involved depended on the physical relationship between the input variable and the output.

The linear relationships between the observed and the predicted sediment discharges by ANNs are better than those by MLR models (Fig. 2). Furthermore, ANNs generate more reasonable predictions for the points with low values where MLP models may give negative predictions, due to the nonlinear transformation process involved (Fig. 2).

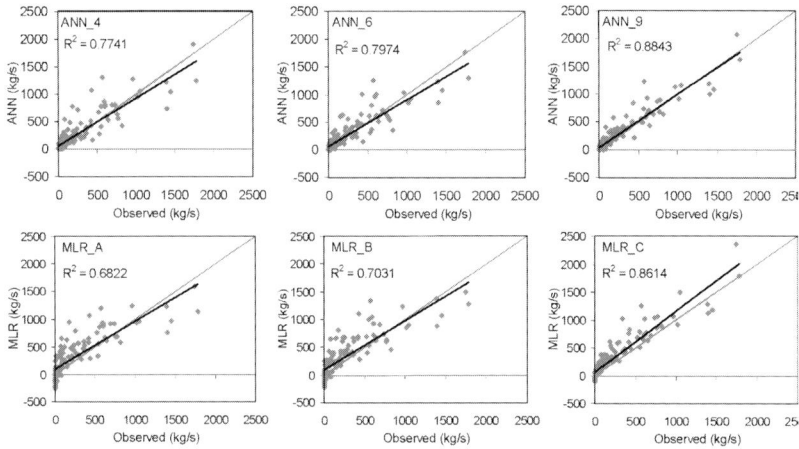

Fig. 2. Scatter plots of the observed and predicted sediment discharges by the corresponding ANNs and MLR models.

5. Conclusion

This study demonstrated that ANN is capable of modeling the monthly sediment discharge with fairly good accuracy when proper variables that represent the driving forces and their lag effect on sediment discharge are used as inputs. The network using both climate variables and water discharge as inputs can provide best simulation; whereas the network with only climate variables as inputs could be used for climate change impact assessment. Compared to the multiple regression models, ANNs can produce better fits to the observed sediment discharge and provide more reasonable results at the extreme points.

The previous researches using ANNs to model water and sediment discharge employed the independent variable at previous time steps as inputs[13] or as the only type of input to the network.[5,9] The ANNs established in this research with only climate variables as inputs have the potential of being used to fill the missing data in sediment time series and to predict the influence of climatic change on sediment discharge.

Acknowledgments

This project is funded by National Basic Research Program of China (Project No. 2003CB415105-6) and National University of Singapore

(Project No. R-109-000-054-112). Acknowledge also goes to the anonymous reviewers for their constructive comments.

References

1. B. Müller, J. Reinhardt and M. T. Strickland, *Neural Networks: An Introduction*. (Springer-Verlag, New York, 1995).
2. ASCE, Artificial neural networks in hydrology. 1: preliminary concepts, *Journal of Hydrologic Engineering* **5**, 2 (2002) 115–123.
3. ASCE, Artificial neural networks in hydrology. 2: hydrology applications, *Journal of Hydrologic Engineering* **5**, 2 (2002) 124–136.
4. M. P. Rajurkar, U. C. Kothyari and U. C. Chaube, Modeling of the daily rainfall-runoff relationship with artificial neural network, *Journal of Hydrology* **285** (2004) 96–113.
5. H. K. Cigizoglu, Estimation, forecasting and extrapolation of river flows by artificial neural networks, *Hydrological Sciences Journal* **48**, 3 (2003) 349–361.
6. R. J. Abrahart and S. M. White, Modelling sediment transfer in Malawe: comparing backpropagation neural netwrook solution against a multiple linear regression benchmark using small data sets, *Physics and Chemistry of Earth (B)* **26**, 1 (2001) 19–24.
7. S. K. Jain, Development of integrated sediment rating curves using ANNs, *Journal of Hydraulic Engineering* **127**, 1 (2001) 30–37.
8. G. Tayfur, Artificial neural networks for sheet sediment transport, *Hydrological Sciences Journal* **47**, 6 (2002) 879–892.
9. Ö. Kisi, Multi-layer perceptrons with Levenberg-Marquardt training algorithm for suspended sediment concentration prediction and estimation, *Hydrological Science Journal* **49**, 6 (2004) 1025–1040.
10. T. E. Croley and H. C. Hartmann, Resolving Thiessen polygons, *Journal of Hydrology* **76** (1985) 363–379.
11. C. W. Dawson and R. L. Wilby, Hydrological modelling using artificial neural networks, *Progress in Physical Geography* **25**, 1 (2001) 80–108.
12. A. S. Tokar and P. A. Johnson, Rainfall-runoff modeling using artificial neural networks, *Journal of Hydrologic Engineering* **4**, 3 (1999) 232–239.
13. S. Raid and J. Mania, Rainfall-runoff model using an artificial neural network approach, *Mathematical and Computer Modelling* **40** (2004) 839–846.

IMPACT OF FLOODPLAIN VEGETATION ON THE SHEAR LAYER AT THE INTERFACE IN 1D MODEL FOR COMPOUND OPEN-CHANNEL FLOWS

SUNG-UK CHOI*, MOONHYEONG PARK and HYEONGSIK KANG

School of Civil and Environmental Engineering
Yonsei University, Seoul 120-749, Korea
*schoi@yonsei.ac.kr

This paper investigates the impact of floodplain vegetation on the shear layer in the one-dimensional (1D) apparent shear stress model for compound open-channel flows. To obtain the mean flow and turbulence structures of the compound channel flows, the three-dimensional (3D) Reynolds stress model is used. The friction slope and interfacial eddy viscosity due to interfacial shear are evaluated and compared with values suggested previously. The impact of proposed interfacial eddy viscosity on the backwater profiles for compound channel flows with and without vegetation on the floodplains are also discussed.

1. Introduction

In general, the straightened wide rectangular channel could cause channel instability, erosion and sedimentation problems, and reduce retention times. Moreover, it can totally isolate the channel flora and fauna from the stream corridor and reduce the natural development of biodiversity.[1] In such cases, an irregular and meandering channel with vegetated floodplain may be an effective low-cost solution.

Hydraulically, floodplain vegetation changes the resistance of the streams, which can affect the flow structure seriously. The difference in the streamwise mean velocities between the main channel and floodplain creates a shear layer at the juncture. The shear layer is related with the twin vortices at the juncture. Through this interface, momentum, and mass are exchanged between the main channel and floodplain. Here, the momentum exchange always exists although net flow carried by each portion remains the same.

The objective of the present paper is to investigate the impact of floodplain vegetation on the interfacial shear layer in the one-dimensional (1D) computation of compound channel flows. For this, a 1D model based on the Apparent Shear stress Method (ASM) is developed. For compound channel flows with and without floodplain vegetation, the friction slope due to interfacial shear and eddy viscosity are evaluated by using the flow structure simulated from the three-dimensional (3D) numerical model developed by Kang.[2] Finally, the impact on the backwater computations is discussed.

2. 1D Mathematical Model

Yen et al.[3] proposed backwater equations for compound open-channel flows, which take into account the flow exchange between main channel and floodplain as well as the shear force between them. For steady and gradually varied flows in compound open-channel, the respective backwater equation for the main channel (subscript m) and the vegetated floodplain (subscript f) can be written as

$$\frac{\mathrm{d}H_m}{\mathrm{d}x} = \frac{S_0 - \left(S_{fm} - \sum_{L,R} S_{sm} + \sum_{L,R} S_{em}\right)}{1 - Fr_m^2}, \tag{1}$$

$$\frac{\mathrm{d}H_f}{\mathrm{d}x} = \frac{S_0 - (S_{ff} - S_{sf} - S_{ef} + S_v)}{1 - Fr_f^2}, \tag{2}$$

where x is the longitudinal distance, H the flow depth, Fr the Froude number, S_0 the bottom slope, S_f the friction slope, S_s the friction slope due to interfacial shear, S_e the friction slope due to flow exchange, and S_v is the friction slope due to vegetation. The friction slope due to vegetation in the floodplain is expressed as

$$S_v = \frac{1}{2}\beta \bar{c}_D a Fr_f^2 h_p, \tag{3}$$

where β is the momentum correction factor, $\bar{c}_D$ the volume averaged drag coefficient of cylinder, h_p the vegetation height, and a is the vegetation density of unit $[\mathrm{L}^{-1}]$. Detailed procedures to obtain S_f, S_s, and S_e are found in Ref. 3.

3. Impact of Floodplain Vegetation

We applied the 3D model by Kang[2] to Tominaga and Nezu's[4] experiment. The flow depths in the main channel and floodplain are 0.08 and 0.04 m, respectively. The width of the main channel (B_m) is 0.2 m, which is the same as that of the floodplain (B_f). The discharge, $Q = 0.0088\,\mathrm{m}^3/\mathrm{s}$, $S_0 = 0.00064$, and the resulting Reynolds number is 54,500.

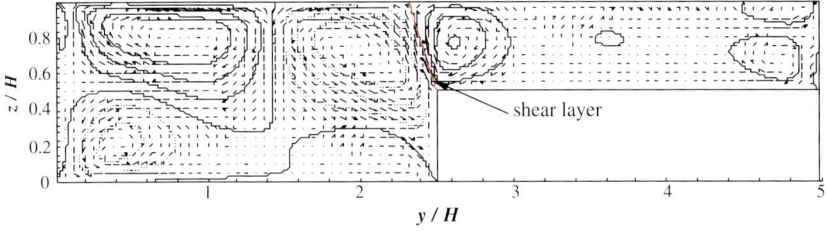

Fig. 1.   3D simulation result of secondary current vectors.

Figure 1 shows the simulated secondary flow vectors and vortical structure. In the figure, a pair of counter-rotating vortices is found at the juncture
between main channel and floodplain. In order to evaluate the friction slope
due to interfacial shear S_s, we define the shear layer along the interfacial
line of twin vortices at the juncture in Fig. 1. The friction slope due to
interfacial shear is evaluated by

$$S_s = \frac{1}{gA_i} \int_s \overline{u'_{\theta s} u'_{\theta n}} \, \mathrm{d}s = \frac{1}{gA_i} \int_s \left(\sin\theta \cdot \overline{u'v'} - \cos\theta \cdot \overline{u'w'} \right) \mathrm{d}s, \qquad (4)$$

where $u'_{\theta s}$ and $u'_{\theta n}$ are the fluctuating velocity components tangent and
normal to the interfacial line, respectively.

4. Results and Discussions

Figure 2 shows the friction slope due to interfacial shear layer for various
flow depth ratios. It can be found that the general trend observed in our
study is the same as that observed in Refs. 3 or 5. Specifically, the friction
slope by the present study lies between their results. Figure 3 shows the
interfacial eddy viscosity estimated by the present study for various flow
depth ratios. The values are of the same order of magnitude as the maximum
value ($= 0.0001$) assumed by Yen *et al.*,[3] but slightly larger than their value.

A backwater computation is performed using Fig. 3. For the com-
putation, such channel data are used as $B_m = 0.28\,\mathrm{m}$, $B_f = 0.32\,\mathrm{m}$,
$H_m - H_f = 0.18\,\mathrm{m}$, $S_0 = 0.00025$, and the roughness coefficients in main
channel and floodplain $n_m = 0.014$ and $n_f = 0.027$, respectively. The dis-
charge of $Q = 0.15\,\mathrm{m}^3/\mathrm{s}$ and flow depth of $H_m = 1.422\,\mathrm{m}$ are imposed at
the upstream and downstream boundaries, respectively. Figure 4 shows the
computed discharge carried by the floodplain along the distance from the

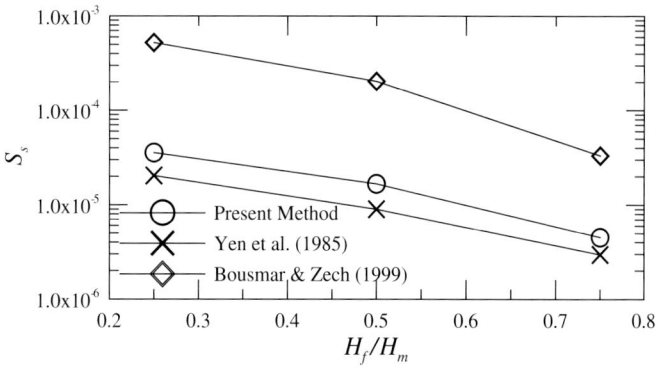

Fig. 2.   Friction slope due to interfacial shear force for plain compound channel flow.

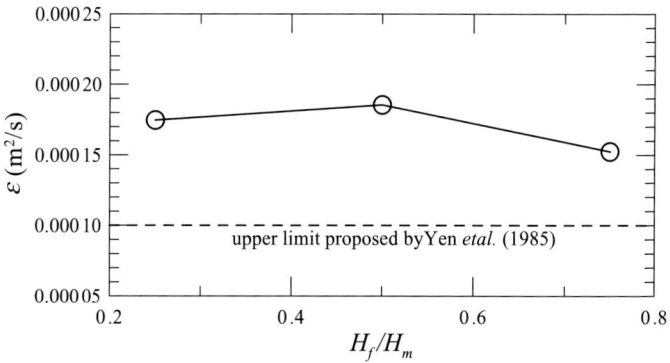

Fig. 3.   Interfacial eddy viscosity.

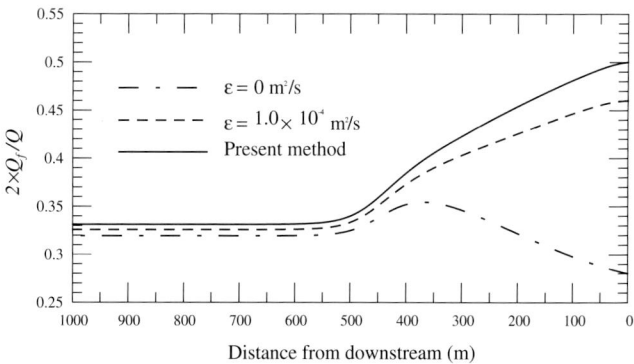

Fig. 4.   Discharge in the floodplains.

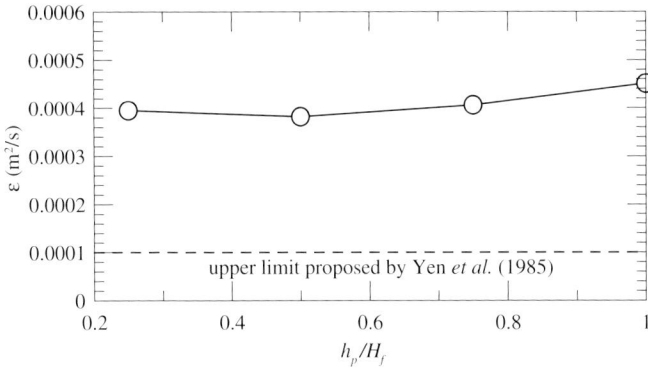

Fig. 5.   Interfacial eddy viscosity versus flow depth ratio.

upstream. It is observed, especially in the backwater region, that the discharge computed by the present method is larger than those by the divided channel method ($\varepsilon = 0$) and by Yen *et al.*[3]

Now, numerical experiments are performed to see the impact of floodplain vegetation on the interfacial shear. A condition of submerged vegetation at vegetation density of $1.0\,\mathrm{m}^{-1}$ is assumed. The results are given in Fig. 5, where the evaluated values of interfacial eddy viscosity are very uniform regardless of the flow depth ratio. It can be seen in the figure that the interfacial eddy viscosity is nearly constant (about 0.0004) irrespective of the flow depth ratio.

A similar backwater computation is carried out for a compound open-channel flow with vegetated floodplain. The vegetation density on the floodplain is assumed to be $a = 1.0\,\mathrm{m}^{-1}$ with $h_p = 0.3\,\mathrm{m}$ and $\bar{c}_D = 1.0$. Other computational conditions are the same as before. The discharge conveyed by the floodplains is given in Fig. 6. A similar result is obtained that the higher value of interfacial eddy viscosity results in larger discharge in both uniform and nonuniform flow regions.

5. Conclusions

This paper investigated the impact of floodplain vegetation on the shear layer in 1D apparent shear stress model for compound open-channel flows. The mean flow and turbulence structures of the compound channel flows were obtained using the 3D Reynolds stress model. It was found that the evaluated friction slope due to interfacial shear is in good agreement with

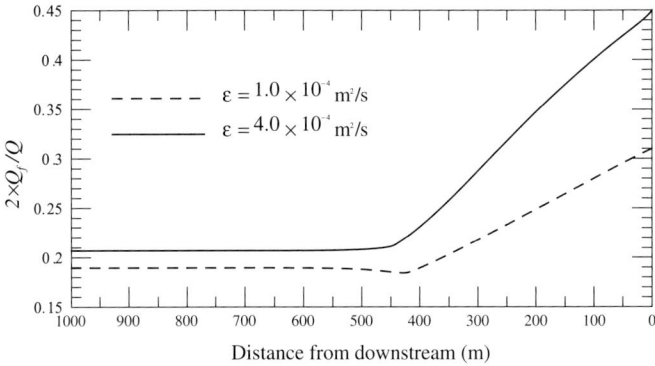

Fig. 6.   Discharge in the vegetated floodplains.

those proposed previously. The interfacial eddy viscosity was estimated, and its impact on the backwater computations was investigated. It was demonstrated that floodplain vegetation strengthens the shear layer at the juncture and reduces the discharge in the floodplains. The method proposed herein suggested that floodplain vegetation further increases the discharge in the floodplains, but decreases the discharge in the main channel, compared with conventional methods.

Acknowledgment

This study was supported by the 2003 Core Construction Technology Development Project (03-SANHAKYOUN-C03-01) through the Urban Flood Disaster Management Research Center in KICTTEP of MOCT KOREA.

References

1. T. Helmio, *Journal of Hydraology* **269** (2002) 89–99.
2. H. Kang, Reynolds stress modeling of vegetated open-channel flows, Ph.D. Thesis, Yonsei University, Seoul, Korea, 2005.
3. B. C. Yen, R. Camacho, R. Kohane and B. Westrich, *Proceeding of the 21st IAHR Congress*, Melbourne, Australia **3** (1985) 439–445.
4. A. Tominaga and I. Nezu, *Journal of Hydraulic Engineering* **117**, 1 (1991) 21–41.
5. D. Bousmar and Y. Zech, *Journal of Hydraulic Engineering* **125**, 7 (1999) 696–706.

OXYGEN TRANSFER BY FLOW CHARACTERISTICS AT STEPPED DROP STRUCTURE

KIM JIN-HONG

Department of Civil Engineering, Chung-Ang University
72-1 Nae-ri, Daeduck-myeon, Ansong-si, Kyeonggi-do, Korea
jinhkim@cau.ac.kr

This paper deals with oxygen transfer by air entrainment by flow characteristics at the stepped drop structure. Nappe flow occurred at low flow rates and for relatively large step height. Dominant flow features included an air pocket, a free-falling nappe impact and a subsequent hydraulic jump on the downstream step. Skimming flow occurred at larger flow rates with formation of recirculating vortices between the main flow and the step corners. Oxygen transfer was found to be proportional to the flow velocity, the flow discharge, and the Froude number. It was more related to the flow discharge than to the Froude number. The stepped drop structure was found to be efficient for water treatment associated with substantial air entrainment.

1. Introduction

Drop structure is useful for air entrainment and energy dissipation by the stepped type of the downstream part of the flow section. Air entrainment by macro-roughness is efficient in water treatment[1] because of the strong turbulent mixing associated with substantial air entrainment. It will be built along polluted and eutrophic streams to control the water quality, since it is used in water treatment for reoxygenation, denitrification and volatile organic component (VOC) removals.[2]

The flows over the stepped drop structures are characterized by the large amount of self-entrained air. The macro-roughness of the steps leads to a sharp increase in the thickness of the turbulent boundary layer. Where the boundary layer reaches the free surface, air is entrained at the so-called inception point of air entrainment. Dissolved oxygen will be made and with this effect, plenty of algae, aquatic insects and fishes can inhabit downstream of the stepped drop structures.

Recently, river environmental works considering ecological habitats have been thought to be important, and the hydraulic studies on the ecological features of the stepped drop structures for the effective design of the river environmental works must be necessary.

The present study deals with the oxygen transfer by the air entrainment by the flow characteristics at the stepped drop structure. Hydraulic analysis on the oxygen transfer by the nappe flow and the skimming flow, and the relationships of the oxygen transfer to the hydraulic parameters were presented through the hydraulic experiments.

2. Flow Characteristics and Oxygen Transfer

Stepped drop structures may be characterized by two types of flows: nappe flow and skimming flow shown in Fig. 1.[1]

At low flow rates and for relatively large step height, nappe flow occurs. The water bounces from one step onto the next one. Dominant flow features include, at each drop, an enclosed air cavity, a free-falling jet, a nappe impact and a subsequent hydraulic jump on the downstream step. Energy

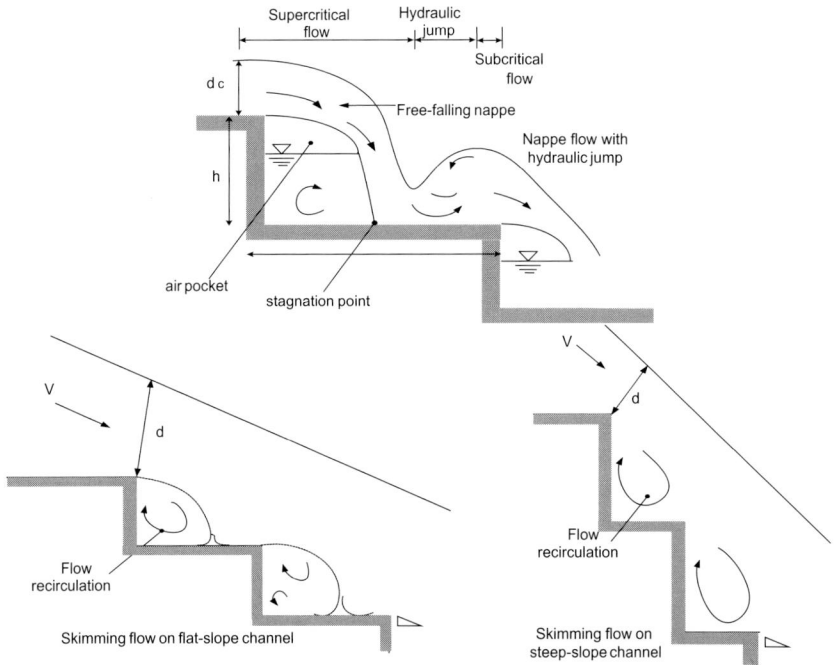

Fig. 1.   Sketch of the nappe and skimming flow.

dissipation takes place due to nappe impact on the underlying water cushion and hydraulic jump.[3]

At larger flow rates and for relatively steep chute, skimming flow occurs. The flow skims over the step edges with formation of recirculating vortices between the main stream and the step corners. The water flows down in a coherent stream where external edges determine a pseudo-bottom defined by the straight line that connects the edges of each step. Most energy is dissipated in maintaining the recirculation in the step cavities.[1]

Oxygen transfer by air entrainment occurs mainly from behind the trailing edge of the stepped structures due to flow separation.[4] Abundant dissolved oxygen is stored with breaking of the air bubbles, and this would give the good habitat condition at the downstream part of the structure. Hydraulic jump makes the air entrainment more active. Occurrence interval of the air entrainment increases when the flow becomes supercritical with Froude number larger than unity.

The efficiency of the oxygen transfer E is used for representing the efficiency of the air entrainment[5];

$$E = (C_d - C_u)/(C_s - C_u),\tag{1}$$

where C_d and C_u are the dissolved oxygen measured at downstream and upstream point, respectively, and C_s is the saturated dissolved oxygen. Since the oxygen transfer is affected by the water temperature, E is substituted by E_{20}[6];

$$\frac{\ln(1 - E_T)}{\ln(1 - E_{20})} = 1.0 + \alpha(T - 20) + \beta(T - 20)^2,\tag{2}$$

where E_T and E_{20} are the oxygen transfer efficiencies at temperature $T°$C and the reference temperature $20°$C, respectively. α and β are constants as $\alpha = 0.02103°$C^{-1}, $\beta = 8.621 \times 10^{-5°}$C^{-2}.

Efficiencies of the oxygen transfer are estimated by measuring the dissolved oxygen and the total head at the upstream and downstream point of the structures, respectively. Figure 2 shows the example for measuring points of the drop structure.

All the data are measured $5\,$m upstream and $10\,$m downstream of the structures for considering the data consistency.

3. Experimental Tests

To investigate the air entrainment by the flow characteristics and to estimate the relationships between the air entrainment and the hydraulic

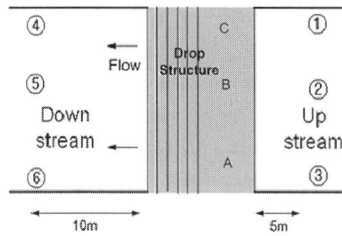

Fig. 2.   Plan and measuring points at the drop structure.

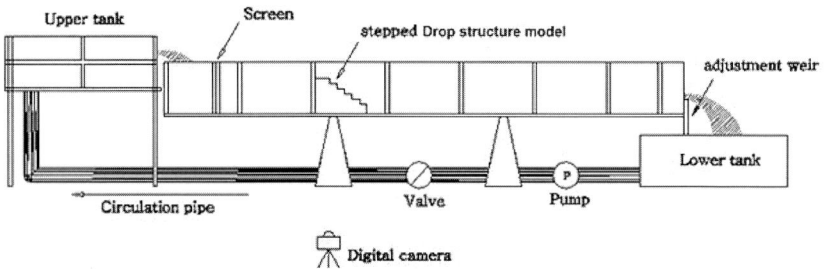

Fig. 3.   Experimental arrangement.

parameters of the stepped drop structure, the experimental tests were performed. Figure 3 shows the experimental arrangements. The typical model of the stepped drop structure made of waterproof plywood was installed in a recirculatory tilting flume of 0.4 m wide, 0.4 m deep and 15 m long. The sidewall of the flume was made of glass and a transparent scale was attached to the side wall to see the flow features well. A damper was laid at the upstream section of the flume to reduce the turbulence and to assure the hydraulic feed having negligible kinetic components. Water level was regulated by the down-stream adjustment weir. The discharge which was controlled by a valve in a feedback loop could be measured with a v-notch at the upper tank.

The stepped drop model was 0.4 m wide and 0.31 m high, and five different slopes were selected (1:2.0, 1:1.7, 1:1.5, 1:1.2, and 1:0.7). Hence, in case of the drop model 1:2.0, the model was 0.4 m wide, 0.54 m long, 0.31 m high and on a slope of 30°. The number of steps was 12, each step was 0.4 m wide, 0.09 m long and 0.052 m high. Flow velocity was measured by using an electromagnetic current meter (model; MI-ECM4). To check the flow pattern, dye injection and a digital camera (model; Olympus c-5050z) with a strong light were used.

4. Results and Discussions

Flow regimes of nappe and skimming flow are shown in Figs. 4 and 5, respectively. Nappe flow occurs at low flow rates and for relatively large step height. Dominant flow features include an enclosed air pocket, a free-falling nappe impact and subsequent hydraulic jump on the downstream step. Air inception occurs from the step edge, but most air is entrained through a free-falling nappe impact and hydraulic jump. Air pocket also has an important role to the air entrainment. The flow accelerated in the downstream direction until a deflected nappe took place. At take-off, free surface aeration was observed at both upper and lower nappes with additional air entrainment at the impact followed by jet breakup.

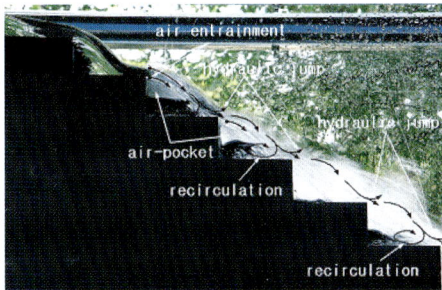

Fig. 4.   Nappe flow.

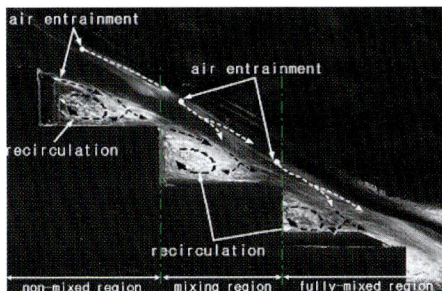

Fig. 5.   Skimming flow.

At larger flow rates, skimming flow occurs with formation of recirculating vortices between the main flow and the step corners. Air entrainment occurs from the step edges. Downstream of the inception point, the flow was highly aerated at each and every step with very significant splashing.

The flow direction of air–water mixture is almost parallel to the pseudo-bottom formed by the step edges although shapes of the recirculating vortices beneath the main flow alternate from step to step. Vortex begins at the upper step and becomes developed at the subsequent downstream step. At the stage of developing vortex, vortex formation is not clear and unstable. Two or three vortices occur and disappear reciprocally. A smaller vortex near the step corner is generated with flow direction opposite to the larger one. Vortex formation is clear and stable at the stage of developed vortex.

Air entrainment was occurred mainly from behind the trailing edge of the drop structure due to flow separation. Air bubbles were formed and proceed to downward direction becoming larger in volume, and finally become broken and disappeared during proceeding upward. Abundant dissolved oxygen was stored with breaking of the air bubbles and this would give the good habitat condition, which is the same ecological feature as riparian riffles.[4]

Figure 6 shows the relationship between the oxygen transfer by flow characteristics. Flow condition changes from a nappe flow to a skimming flow as the flow velocity and Froude number increase. The transition between nappe and skimming flow was shown to occur at region of $v = 0.56$–$0.79\,(\text{m/s})$ and $Fr = 1.32$–1.51. This was due to the undular profile of the free surface, acceleration above filled cavities and deceleration at nappe impact as was suggested by Chanson and Toombes.[7]

Oxygen transfer becomes smaller and reaches to minimum value at the beginning stage of a skimming flow, but becomes larger in the region of

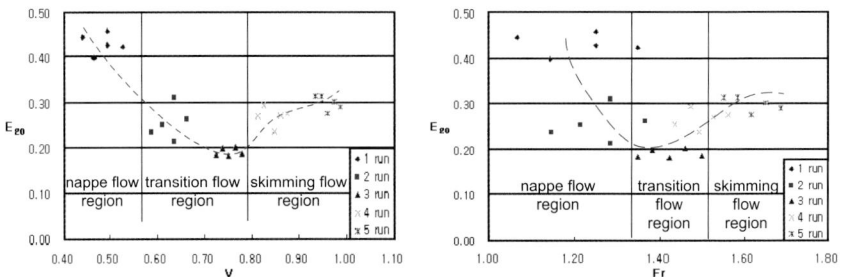

Fig. 6.   Relationship between oxygen transfer and flow parameters.

skimming flow because air entrainment is made mainly through a free-falling nappe impact, a hydraulic jump and an air pocket in the region of nappe flow. The average values of the oxygen transfer efficiency in the region of the nappe flow and in the region of the skimming flow are about 0.45 and 0.28, respectively. The stepped drop structure is found to be efficient for water treatment associated with substantial air entrainment.

Acknowledgments

This research was supported by the Chung-Ang University Research Grants in 2004.

References

1. H. Chanson, Self-aerated flows on chute and spillways, *Journal of the Hydraulics Division*, ASCE **119**, 2 (1993) 220–243.
2. T. Henry, *Air-Water Flow in Hydraulic Structures*. A Water Resources Technical Publication, Engineering Monograph No. 41 (1985), pp. 251–285.
3. U. Fratino, and A. E. Piccinni, Dissipation efficiency of stepped spillways, *Proceedings of the Nineth International Workshop on Hydraulics of Stepped Spillway*, Zurich, Switzerland (2000), pp. 103–110.
4. J. H. Kim, Water quality management by stepped overflow weir as a method of instream flow solution, *Proceedings of the First International Conference on Solutions of Water Shortage and Instream Flow Problems in Asia*. Incheon, Korea (2003), pp. 24–36.
5. S. T. Avery and P. Novak, Oxygen transfer at hydraulic structures, *Journal of the Hydraulics Division*, ASCE **104**, 11 (1978) 1521–1540.
6. J. S. Gulliver, J. R. Thene and A. J. Rindels, Indexing gas transfer in self-aerated flows, *Journal of the Environmental Engineering*, ASCE **116**, 3 (1990) 503–523.
7. H. Chanson and L. Toombes, Hydraulics of stepped chutes: The transition flow, *Journal of the Hydraulic Research*, IAHR **42**, 1 (2004) 43–54.

IMPACTS OF FOREST DEFOLIATION BY PINE-WILT DISEASE ON BIOGEOCHEMICAL CYCLING AND STREAMWATER CHEMISTRY IN A HEADWATER CATCHMENT IN CENTRAL JAPAN

NOBUHITO OHTE[*,†], NAOKO TOKUCHI[§] and SATORU HOBARA[‡]

*Graduate School of Agriculture, Kyoto University
Kyoto 606-8502, Japan
†nobu@bluemoon.kais.kyoto-u.ac.jp

§Field Science Education and Research Center, Kyoto University
Kyoto 606-8502, Japan

‡Faculty of Environmental Systems, Rakuno Gakuen University
Ebetsu 069-8501, Japan

In order to evaluate the impact of the forest disturbance by pine-wilt disease (PWD), changes in nutrient status in soils and streamwater chemistry have been investigated in a small headwater catchment in central Japan, focusing on the nitrogen dynamics and hydrological processes. Decreased N uptake by roots and increased N supply from litter fall caused by the 1992–1994 PWD caused a threefold increase in NO_3^- and Ca^{2+}, and Mg^{2+} concentrations of stream and groundwater. Seasonal peaks in stream NO_3^- concentration during the rainy season occurred during 1992–1996. The mechanism of this seasonal pattern can emphasize the importance of hydrological seasonality with high precipitation, groundwater level in summer of Japan under Asian monsoon climate. The N loss through the streamwater was much smaller than N contribution of PWD litter inputs throughout the observation period. This large discrepancy suggested substantial nitrogen immobilization in soils.

1. Introduction

Several forest declines have been reported in Japan since the 1960s. A major disaster for the Japanese forestry industry has been the decline of Japanese pine forests due to pine-wilt disease (PWD), which over the past three decades has spread extensively in western Japan.

The mechanisms of forest decline due to PWD have been studied continuously since the 1970s.[1] PWD is caused by pinewood nematode (*Bursaphelenchus xylophilus (Steiner et Buhrer) Nickle*[2]), and its vector is a sawyer beetle (*Monochamus alternatus*[3]). Recently, this forest declines have also become a major concern for Japanese citizens utilizing water resources from forested catchments, since there is a possibility that severe disruption of the nutrient cycle of a forest ecosystem will result in deterioration in drainage water chemistry, such as unusual elevation of stream NO_3^- concentrations or eutrophication of lake water.

Biogeochemical dynamics have been investigated in tandem with groundwater monitoring in a temperate forest catchment in Japan, to determine the effect of the partial dieback of dominant tree stands on nutrient cycling and streamwater chemistry, especially temporal and spatial changes in the solute concentrations related to nitrogen dynamics in soil and ground water.[4]

2. Field Investigation

The field investigation has been carried out in a forested headwater catchment (0.68 ha, the Matsuzawa catchment) in the Kiryu Experimental Watershed in central Japan (35°N, 136°E). The catchment is underlain by weathered granitic rocks. The soils are predominantly cambisols. Detail site descriptions were in Refs. 4 and 5 and Fig. 1.

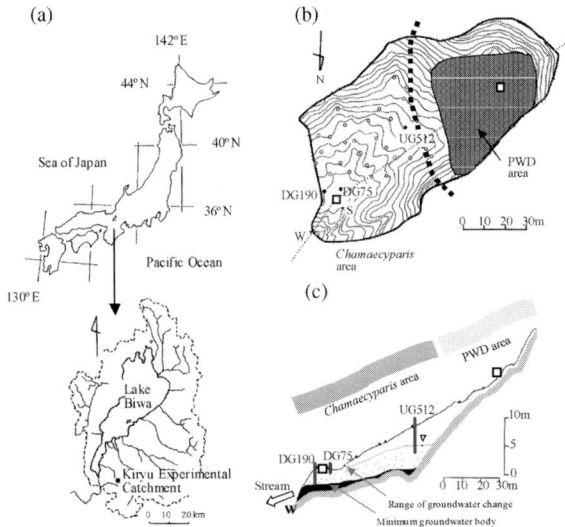

Fig. 1.   (a) Location of the Kiryu Experimental Watershed. (b) Surface topography and observation apparatus of the Matsuzawa. UG512, DG75, and DG190 indicate the observation wells. The two solid squares indicate points where soil solutions were sampled. S: spring point; W: gauging weir; Open circles indicate other wells. (c) Longitudinal section along the dotted line in the panel (b) and groundwater condition (after Ohte *et al.*[4]).

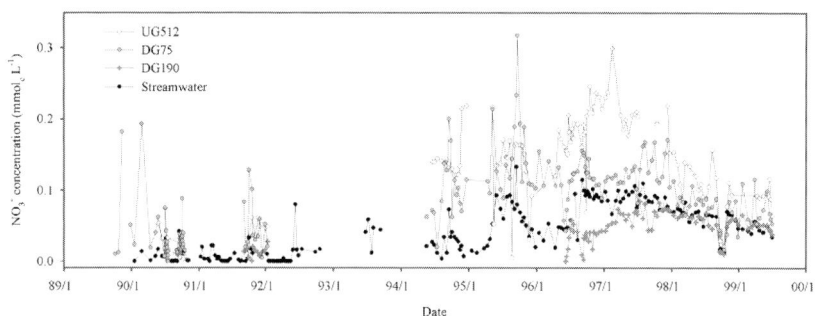

Fig. 2. Changes in the NO_3^- concentrations in the stream and in the groundwater at UG512, DG75, and DG190 from 1989 to 1999. The well codes and locations are shown in Fig. 1 (after Ohte *et al.*[4]).

The catchment was covered by semimature vegetation (a mixed stand of *Pinus densiflora* and *Chamaecyparis obtusa*) before the PWD diebacks of the early 1990s. Over 25% of the Matsuzawa catchment was suffered by PWD (Fig. 1a). The total basal area of *Pinus* in the PWD portion decreased from 18.6 to $1.9 \, m^2 \, ha^{-1}$ from 1989 to 1998.[6]

Sampling design of waters were described in Ref. 4. Three observation wells were installed in the catchment: two were located at the furthest downstream reach of the groundwater body (DG75, DG190), and one at the furthest upstream reach of the groundwater body (UG512, see Fig. 1b). The respective depths of the well bottom and the bedrock are 75 and 445 cm for DG75, 190 and 450 cm for DG190 and 512 and more than 800 cm for UG512. It should be noted that the water sampled from DG190 could include relatively deeper groundwater than samples taken from DG75.

Visible dieback of pine stands has been observed since 1990.[6] Decreased N uptake by roots and increased N supply from litter fall caused by the 1992–1994 pine dieback caused a threefold increase in NO_3^- and cations (Ca^{2+} and Mg^{2+}) concentrations of streamwater and subsurface groundwater.[7] Unusual increase in NO_3^- concentration of streamwater occurred since 1994. From 1992 to 1996, remarkable seasonal peaks was observed in the stream NO_3^- concentrations during the rainy season from July to August (Fig. 2)[4].

3. Alterations in Nutrient Status by PWD

It was estimated that leaf litter during PWD supplied nitrogen and calcium of $0.17 \, kmol \, ha^{-1}$ and $0.08 \, keq \, ha^{-1}$ respectively, and tree branches and stems supplied more than $7.3 \, \mathrm{kmol \, ha^{-1}}$ and $3.7 \, \mathrm{keq \, ha^{-1}}$ of nitrogen and calcium, respectively.[7] In 1989, there was no apparent difference in soil nitrogen mineralization rates between the pre-PWD site and the SZ (nondefoliation site surrounding DG75): 0.6–$7.5 \, \mathrm{mgN \, kg^{-1}}$ at the PWD site versus 0.6–$9.1 \, \mathrm{mgN \, kg^{-1}}$ at the SZ site in 28 days (Fig. 3a and d). There was little nitrification at either site; mineralized nitrogen remained as NH_4^+. In 1991, soil nitrogen mineralization rates at both sites were accelerated (Fig. 3b and e[7]).

At the *Chamaecyparis* site (surrounding DG75), the high mineralization rate was not constant, whereas at the PWD site it was still high in 1997. The mineralization rate on the soil surface at the PWD site was three times higher in 1997 than in 1989 (Fig. 3c). It was significantly higher than at the *Chamaecyparis* site ($p = 0.04$, Fig. 3f). In 1997, more than 90% of mineralized nitrogen was nitrified at the PWD site. The ratio of nitrification to mineralization was significantly larger at the PWD site than at the *Chamaecyparis* site ($p = 0.04$, Fig. 3c and f[7]).

It was estimated that dry matter of approximately $82.8 \, \mathrm{Mg \, ha^{-1}}$ and nitrogen at $7.39 \, \mathrm{kmol \, ha^{-1}}$ were added to the forest floor from tree die back with PWD, which were three to five times larger than that due to litter fall. Increase of net nitrogen loss via the stream was much smaller than the nitrogen addition by PWD, although the stream NO_3^- concentration increased. Vegetation uptake (regeneration) at the PWD site was important for considering the retention mechanism of nitrogen. Alternatively, microbial nitrogen immobilization could have affected the nitrogen budget, since the PWD litter, including stems and branches, had a relatively high C/N ratio.[8]

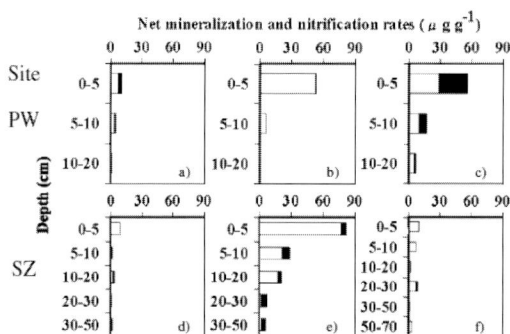

Fig. 3. Nitrogen transformation rates through the soil profile of SZ (a–c) and PW (d–f) sites pre-PWD (1989; a and d), during-PWD (1991; b and e), and post-PWD (1997; c and f). Black indicates NO_3^-; white shows NH_4^+.[7]

4. Discharge Mechanisms of NO_3^-

NO_3^- leaching from the near-surface horizon was prominent from the second half of October to the mid-November in 1997. In 1998, NO_3^- levels in soil water began to increase in mid-August.[4] In association with the vertical infiltration of soil water, NO_3^- was transported from near the surface to the lower layers with dispersion. The seasonal pattern, however, became unclear compared with concentrations in the shallower layer. It is unlikely that the seasonal signal in soil water directly affected the seasonal variation in the stream NO_3^- concentrations.[4]

The fluctuation in streamwater NO_3^- concentrations corresponded strongly with groundwater levels at DG75, suggesting that volumetric changes in the contribution of shallow groundwater from the DG75 zone affected the seasonal variation in the stream NO_3^-.[4]

5. Conclusions

Although very high potential of N retention has been found in soil bio-geochemical processes, the increase in stream NO_3^- concentration was remarkable. The mechanisms for the seasonal patterns in stream NO_3^- concentrations in this case emphasize the importance of changes in hydrological conditions in summer in Japan, where precipitation, groundwater levels, and runoff rates are high. Previously, in Europe and the eastern United States, the simultaneous effects of smaller pools of inorganic nitrogen due to high plant uptake and lower hydrologic transporting forces due to low precipitation and high transpiration during the summer[9] may let ones underrate the effect of hydrological conditions on seasonality in stream NO_3^- concentrations.[10]

In order to precisely understand the mechanism behind the biogeochemical responses to the catchment scale disturbances, including NO_3^- leaching due to main canopy defoliation, hydrological information are correspondingly necessary.

Acknowledgments

I would like to thank Masanori Katsuyama, Masatoshi Kawasaki, and Keisuke Koba for their critical efforts on the field and laboratory works for this project.

References

1. K. Fukuda, *J. For. Res.* **2** (1997).
2. T. Kiyohara and Y. Tokushige, *J. Jpn. For. Soc.* **53** (1971) (in Japanese with English summary).
3. Y. Mamiya and N. Enda, *Nematology* **18** (1972).
4. N. Ohte, N. Tokuchi, M. Katsuyama, S. Hobara, Y. Asano and K. Koba, *Hydo. Proc.* **17** (2003).
5. N. Ohte, N. Tokuchi and M. Suzuki, *Wat. Resou. Res.* **31** (1995).
6. S. Hobara, N. Tokuchi, N. Ohte, A. Nakanishi, M. Katsuyama and K. Koba, *Can. J. For. Res.* **31** (2001).
7. N. Tokuchi, N. Ohte, S. Hobara, S. Kim and M. Katsuyama, *Hydro. Proc.* **18** (2004).
8. G. M. Lovett, L. M. Christenson, P. M. Groffman, C. G. Jones, J. Hart and M. J. Mitchell, *BioScience* **52** (2002).
9. M. J. Mitchell, G. Iwatsubo, K. Ohrui and Y. Nakagawa, *For. Eco. Man.* **97** (1997).
10. N. Ohte, M. J. Mitchell, H. Shibata, N. Tokuchi, H. Toda and G. Iwatsubo, *Wat. Air, Soil Poll.* **130** (2001).

LESSONS FROM JAPANESE EXPERIENCE IN INVOLUNTARY RESETTLEMENT FOR DAM CONSTRUCTION — CASE OF NEW VILLAGE BUILDING

NARUHIKO TAKESADA

Graduate School of Frontier Sciences, University of Tokyo
Hongo 7-3-1, Bunkyo-ku, Tokyo, Japan
naruhiko@d02.itscom.net

Constructing dams, as one of measure approaches in water resources development in developing countries, often involves number of households to be resettled involuntarily and is nowadays severely criticized. In postwar Japan, there were many cases of dam construction with involuntary resettlement. In this study, involuntary resettlement of Ikawa Dam in Shizuoka Prefecture is examined as a case. Ikawa Dam, completed in 1957 in the Ohi River, was constructed for the purpose of hydro power generation. The special approach of compensation taken instead of monetary compensation for resettled 193 households was "New Village Building." Among several enabling conditions for "New Village Building," the role played by Shizuoka Prefecture government has been focused on. Reviewing these experiences of Ikawa Dam Project will give us valuable insights applicable to the current resettlement practice in developing countries.

1. Introduction

Dams often submerge large areas, where people live, cultivate, or derive necessary resources for their livelihoods, sometimes near a city or even city itself, but usually in rural villages. Persons affected must leave their ancestral lands with compensation and need to start new lives elsewhere either near the original place or far away. In the World Bank estimate in 1994, every 300 large dams which enter into construction in every year result more than four million people displaced. Involuntary resettlement has been and is being reiterated in the world wide scale, both in developed and developing countries.[1]

This "involuntary resettlement" used to be perceived as necessary sacrifice or side-effect for national development or economic growth. However, the risk of impoverishment of these resettled people is widely accepted nowadays. After Narmada development controversy in India in the end

143

of 1980s, NGOs are strongly critical to dam development with involuntary resettlement and mobilize international campaign against such development. On the other hand, states and donors become more cautious to implement such projects by adopting more sophisticated procedures such as improved safeguard policies in order to avoid conflict and impoverishment of resettled people.[2]

2. Dam Development and Involuntary Resettlement in Japan

In Japan, there are more than 2,500 dams constructed. Especially, in 1950s and 1960s, with the growing demand for electricity and water as well as flood control, there was a boom for dam construction. Although, in Japan, the number of households affected in one dam construction were relatively small (around a few hundred) compared with the current developing countries' cases, there were a number of dam constructions with involuntary resettlement. In 1962, "the Guideline on Compensation Standard for the Loss Caused by Public Land Acquisition" ('Sonshitsu Hosho Kijun Yoko' in Japanese) was established and since then compensation has been made in monetary basis commonly in any public works including dam construction. Before having established this guideline, some of the project promoters had devised and tried innovative and unique way of compensation considering reconstruction of livelihood of resettled inhabitants.[3] In developing countries today, reflecting the observation that large amount of money as compensation was spent away immediately or put resettlers into difficulties, reconstruction of livelihood through income restoration is perceived as one of major challenges in involuntary resettlement.[4] Therefore, it is assumed that in Japanese early experiences there might be lessons or knowledge which may be applicable or useful in the current practice in developing countries.

3. Ikawa Dam Construction and Resettlement — New Village Building

Ikawa Dam in the Ohi River, located in Central Japan, was completed in 1957 for hydro electric power generation built by Chubu Electric Power Company. Although the plan to construct Ikawa dam was conceived long back, it had to be waited until 1952 that realization of this plan was firmly and formally put on the agenda. Height of the dam is 103 m. Storage capacity is 125 million m^3. With 422 ha of inundated area, 193 households were relocated among 553 total households in then Ikawa village. Although whole village was not inundated, the center of the main village as well as several hamlets along the river was submerged.

Due to this large (relatively in Japan) number of affected households, negotiation between Chubu Electric Power Company and villagers faced with difficulties. After rigorous consultation among parties, Chubu Electric Power Company finally agreed to three principles of compensation requested by villagers. These are: (1) to complete Dainichi Road (road to Shizuoka city), (2) to build New Village for civilized life, and (3) to compensate fully and satisfactorily for better and improved living standard. One of major villagers' requests was to build "New Village."[5,6] At the same time, there was an option to have cash compensation instead and 99 households among 193 affected households took this option, to leave the village.

This compensation scheme, New Village Building aimed to provide with compensation on land-for-land basis. Some of villagers, who lost houses and cultivated land, obtained new housing and new land plot in a newly developed area in the village. Others received new reclamation land for their housing within the original main village which was also equipped with new infrastructure and community facilities. In one of such newly developed areas, called Nishiyama-daira, near the main village, 23 houses were built with attached land plots and other community facilities including water supply, electricity supply.

There were two special features in this compensation scheme. First, in the newly developed land, rice cultivation was also newly introduced. At high altitude around 700 m, Ikawa village had not had any substantial paddy field before inundation. Their main staple food had been millet, with slash and burn agriculture. It is said that many villagers had known rice as a very precious commodity but not rice plant itself. Second, for this new agricultural practice, one agricultural expert was stationed for 4 years in Nishiyama-daira in order to assist villagers to stabilize their agricultural productions and hence their livelihoods. With a help of the expert, villagers moved to the newly developed land had substantial success in rice farming and their livelihoods after relocation was improved in a very short time. New Village Building could bring stable and secure livelihood to villagers.

4. Analysis on Enabling Conditions

What made this New Village Building enable? From several records and interviews with villagers, involved in the negotiation with Chubu Electric Power Company, following points can be pointed out.

4.1. Enabling conditions for new village building

First, Chubu Electric Power Company was so eager to build Ikawa Dam to supply electricity for regional reconstruction and industrialization that they accepted larger cost for land-for-land compensation than the cost for land-for-money compensation.

Second, villagers had been long aware of the possibility of dam construction in their village. One of villagers, involved in the negotiation, clearly remembers his childhood when survey teams occasionally visited and conducted preliminary survey. Therefore, when the issue was put on the agenda formally, villagers were ready to discuss the request among them and were quite active to do so.

Third, and most importantly, Shizuoka Prefecture Government (local government) played very active role in this compensation scheme. In the next part, further analysis on this point will follow.

4.2. Role played by the local government

Shizuoka Prefecture Government took active role in threefold. First, her basic position was as a *stakeholder* of Ikawa Dam construction. The construction of Ikawa Dam and its hydroelectric power plant was not only corporate target of Chubu Electric Power Company but also a part of Comprehensive Development Plan of Shizuoka Prefecture. For her economic growth with industrialization, electric power supplied by Ikawa Dam was perceived quite essential. Also, in then Japan, the development of major rivers was under the authority of the governor not of the central minister. Therefore, it was also possible for prefecture government to have a genuine stake in the river development within her territory.

Second, she played a role as an *arbitrator* between villagers and Chubu Electric Power Company. It was in the first term of newly elected governor and just 4 years since public election of governor was realized in postwar Japan. Therefore, expectation to the authority of publicly elected governor was high in local development. This expectation was also revealed when the decision on the amount of solatium, which was the last issue undecided between villagers and Chubu Electric Power Company, was entrusted to the governor.

Third role of prefecture government was as a *planner and advisor* of New Village Building. The very idea of New Village Building was devised by the officials of Shizuoka Prefecture. Initially, villagers were skeptical about the plan since they had witnessed previous unsuccessful attempt to develop new land nearby the village in order to accommodate returnees from the World War II. However, after conducting feasibility study of rice cropping including experimental farming and attending the negotiation with villagers, officials promoting the plan gradually got confidence of villagers. In this course of action, one agricultural expert from the prefecture office was stationed in the village for 4 years.

5. Conclusions and Implications for Current Practices in Developing Countries

In Ikawa case, land for land compensation (New Village Building) and active assistance from the local government helped villagers to quickly restore and reconstruct their livelihoods. When reflecting several experiences in developing countries' project with involuntary resettlement, it is found that there are not a few cases of compensation in kind but less involvement and/or coordination with local government in compensation and reconstruction of resettlers' livelihoods. Only coordination among central ministries is usually emphasized (but usually not succeeded) since the project is promoted by the central ministry or agency as a national project.

If coordination with central government agency, and/or active role, is pursued by the local government, in line with general trend of decentralization in developing countries, compensation and reconstruction of resettlers' livelihood may have favorable result. If local government is able to provide resettlers with sustained assistance as in Ikawa case, it may lower the risk of impoverishment, since local government is in a better position to know the local conditions and environment.

This paper does not try to judge resettlement in Ikawa Dam as success or failure. If seeing current situation of Ikawa area after 50 years of resettlement, one may find that the area is now very under populated with more than 50% of villagers older than 65. However, in Nishiyama-daira, there are still 24 households living without decreasing the number of household. Rice cultivation is not pursued any more except by two households while the given plots are utilized for vegetable cultivation as well as tea plantation.

In sum, it is not the matter of success or failure. Still further analysis of consequences of involuntary resettlement in Japan such as Ikawa case is to follow in order to have valuable insights for the path of development, on which developing countries will travel.

Acknowledgments

This research was partly funded by the New Research Initiatives in Humanities and Social Sciences and a Grant-in-Aid for Scientific Research (No. 15510034) of the Japan Society for the Promotion of Science (JSPS).

References

1. The World Bank Environment Department, *Resettlement and Development — The Bank-wide Review of Projects Involving Involuntary Resettlement 1986–1993*, 1/3 (The World Bank, Washington, 1994).
2. R. Dwivedi, Models and methods in development-induced displacement, *Development and Change* **33**, 4 (2002) 709–732.
3. K. Hanayama, *Hosho no Riron to Genjitsu (Theory and Practice in Compensation)* (Keiso Shobo, Tokyo, 1966).
4. The World Bank, *Involuntary Resettlement Sourcebook — Planning and Implementation in Development Projects* (The World Bank, Washington, 2004).
5. Ikawa Village, *Ikawa Dam no Kiroku* (Record of Ikawa Dam, 1958).
6. Chube Electric Power Company Construction Department, *Ikawa Dam Koujisi (Report of Ikawa Dam Construction*, 1961).

INTER AND INTRA NEURONAL SYSTEMS FOR RESERVOIR OPERATION

RAMANI BAI VARADHARAJAN

University of Malaya, Kuala Lumpur, Malaysia

Modeling of the human brain is done at two levels, a macroscopic, inter-neuronal level and a microscopic, intra-neuronal level. With the former, various models have been proposed in the literature for artificial neural networks (ANN) dealing with such specific brain functions as function approximation, pattern recognition, classification and control. In modeling at a microscopic level, energy transfer across the cells without dissipation is conjectured in biological matter. This idea has led to construct models for water reservoir operation, which can assist water resources system analyst in selecting compromise strategies for water release from the reservoir. Multilayer Feed forward networks such as Back-propagation with Levenberg–Marquardt Algorithm (BPLM) and Recurrent Neural Network (RNN) model as inter and intra neuronal architectures are formed. The aim is to find a near global solution to what is typically a highly non-linear optimization problem like reservoir operation. The main reason for this study is due to the research results[1] on back-propagation (BP) technique and its performance. The case study used in this paper to demonstrate the effect of these architecture geometry and various internal parameters is that of deriving operational policy for the river Vaigai in south of Tamil Nadu in India.

1. Introduction

Learning denotes changes in the system that are adaptive in the sense that they enable the system to do the same task or tasks drawn from the same population more effectively next time. The primary objective of the reservoir operation is to maintain operation conditions to achieve the purpose for which it has been created with least interference to the other systems. In this paper, integration of surface and groundwater available is made use of as resource available for water meeting irrigation demand of the basin. Effects of inter basin transfer of water between Periyar and Vaigai system is also carefully studied in determining release policies. The inter-basin transfer of water not having negative environmental impacts is a good concept but that should also have social and cultural consideration, price considerations and some environmental justice considerations. There are significant water-sharing conflicts within agriculture itself, with the various agricultural areas competing for scarce water supplies. Increasing basin water demands are placing additional stresses on the limited water resources and threaten its quality. Many hydrological models have been developed to problem of reservoir operation. System modeling based on conventional mathematical tools is not well suited for dealing with nonlinearity of the system. By contrast, the feed forward neural networks can be easily applied in non-linear optimization problems.

Jiri[2] proposed the back-propagation learning algorithm for multilayered neural networks, which is often successfully used in practice, appears time consuming even for small network architectures or training tasks. Clair and Ehrman[3] used a neural network approach to examine relationships between climate and geography on discharge and Dissolved Organic Carbon (DOC) and Dissolved Organic Nitrogen (DON) from 15 river basins in Canada's Atlantic region over 10 years period. They emphasized the importance of the evapotranspiration–precipitation link in establishing basin discharge, because even large increase in precipitation can lead to decrease the discharge when accompanied by higher temperature. Yang et al.[4] presented a flood forecasting procedure by integrating linear transfer function (LTF), autoregressive integrated moving average (ARIMA) model and ANN. They illustrated the integrated method and a stand-alone ARIMA model applied to Wu-Shi basin, Taiwan. The results obtained from these two models were compared. It was concluded that the integrated ANN model, which consists of ANN, LTF, and ARIMA model is appropriate for watershed flood forecasting. Chandramouli and Raman[5] developed for optimal multireservoir operation, a dynamic programming-based neural network model. The training of the network is done using a supervised learning approach with the back-propagation algorithm. Rules derived for the three reservoirs using dynamic programming-neural network model gave better performance than dynamic programming-regression model. This paper gives a short review on two methods of neural network learning and demonstrates their advantages in real application to Vaigai reservoir system. The developed ANN model is compared with LP model for its performance. This is also accomplished by checking for equity of water released for irrigation purpose. It is concluded that the coupling of optimization and heuristic model seems to be a logical direction in reservoir operation modeling.

2. Study Area and Database

The model is applied to Vaigai basin (N $9°15'$–$10°20'$ and E $77°10'$–$79°15'$) which is located in south of Tamil Nadu in India (Fig. 1). The catchment

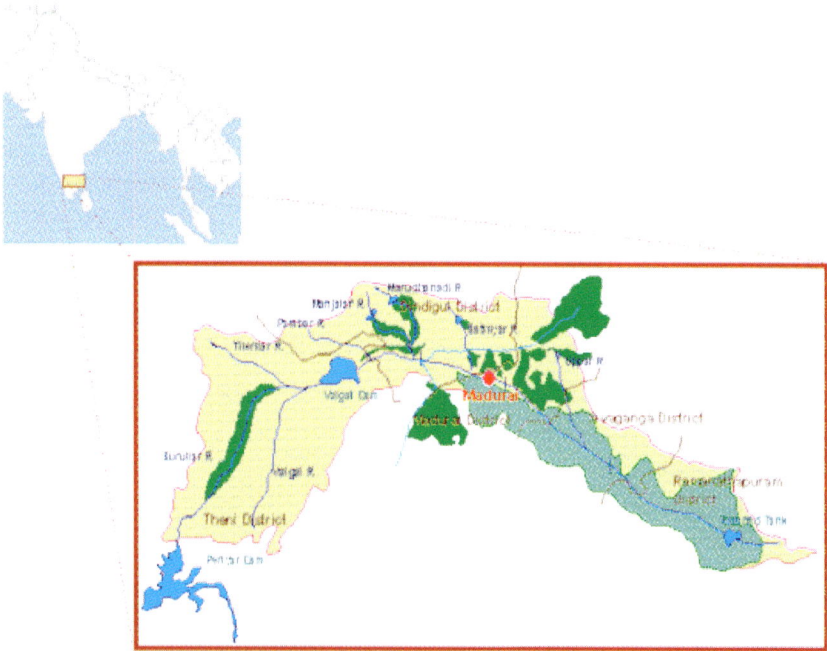

Fig. 1. Location of Vaigai dam in south India.

area is 2,553 sq. km and water spread area is 25.9 sq. km. The reservoir is getting water from catchments of two states namely, Tamil Nadu and Kerala. Maximum storage capacity of the dam is $193.84\,\mathrm{Mm^3}$. The period of study is from 1969 to 1997 of measured historical data. Flow data from water year 1969 to 1993 is used for training of the neural network model for operation of the reservoir and data from water year 1994 to 1997 is used for validation of the model.

3. Model Development

First, a linear programming model is developed with the objective function consists of maximizing the net annual benefit of growing six primary crops in the whole basin deducting the annualized cost of development and supply of irrigation from surface and groundwater resources.

For maximizing net benefit for every subarea,

$$\max \sum_{i=1}^{n} \sum_{c=1}^{6} B_{i,c} A_{i,c}, \qquad (1)$$

where

c = crops, 1–6 (single crop, first crop, second crop paddy, groundnut,
　　 sugarcane, and dry crops),

i = nodes in Zones II–V or subarea,

$B_{i,c}$ = net annual benefit after production cost and water cost for crop c
　　 grown in subarea i in Rs/Mm^2 (Rs = Rupees, Indian Currency),

$A_{i,c}$ = land area under crop c in subarea i in Mm^2.

The accounting system of water, which is still in use in Vaigai system, is taken into consideration in the model with all other constraints as:

1. crop water requirement constraints,
2. land area constraints,
3. surface water availability constraints,
4. groundwater availability constraints,
5. continuity constraints.

The detailed linear programming model has been discussed in Ref. 6. It is an optimization model with in-depth discussion on modeling in an array of application areas using Simplex method. The model is run for 29 years of inflow (1969–1997) into the basin.

Second, an artificial neural network (ANN) structure shown in Fig. 2 can be applied to develop an efficient decision support tool considering the parameters, which are non-linear in nature and to avoid addressing the problem of spatial and temporal variations of input variables. By this work an efficient mapping of non-linear relationship between inflow, storage, demand and release pattern into an ANN model is performed for better prediction on optimal releases from Vaigai dam. The back-propagation network (BPN) is the most popular network among recent applications of ANN.[7] Feed forward Error Back-propagation Network (FFBPN) with Levenberg Marguardt model is considered in this study. The hyperbolic tangent transfer function, log sigmoid transfer function; the normalized cumulative delta learning rule and the standard (quadratic) error function were used in the framework of the model.

$$E(t) = \frac{1}{2} \sum \left(d_j(t) - y_j(t) \right)^2, \qquad (2)$$

where $E(t)$ is the global error function at discrete time t and $y_j(t)$ is the predicted network output at discrete time t and $d_j(t)$ is the desired network output at discrete time t.

Input Layer

Hidden Layer

Output Layer

Tt

It

**Input
Signals**

St

Dt

Rt

**Output
Response**

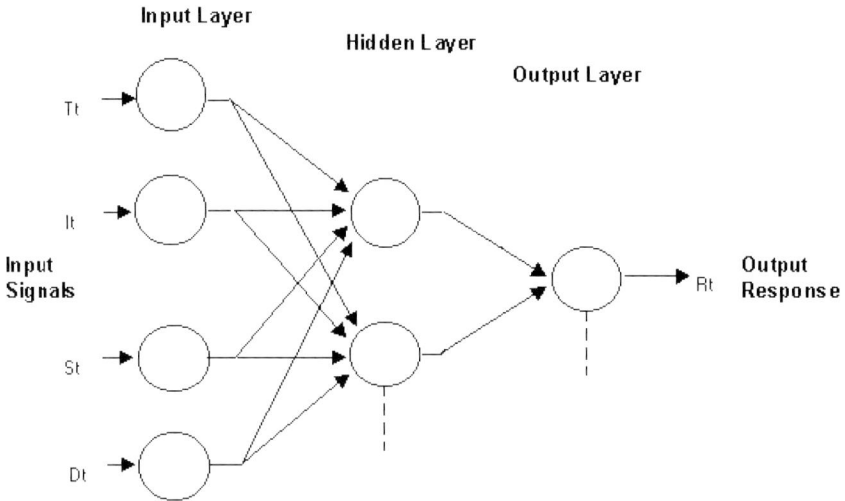

Fig. 2.   Architecture of three-layer FFNN.

This calibration process is generally referred to as "training". The global error function most commonly used is the quadratic (mean-squared error) function.

The connection weights are then adjusted using a form of the generalized delta-learning rule in an attempt to reduce the error function. The amount by which each connection weight is adjusted depends on the learning rate (η), the momentum value (μ), the epoch size ($\in$), the derivative of the transfer function and the node output. The weight update equation for the connection weight between nodes i and j is given in Eq. (3).

$$\Delta w_{ji}(t) = \sum_{s=1}^{e} \{\eta(d_j - y_j)f(\cdot)y_i\} + \mu \Delta w_{ji}(t-1), \qquad (3)$$

where w_{ji} is the connection weight between nodes i and j, $(d_j - y_j)$ is the difference between actual and predicted values (error), $f(\cdot)$ is the derivative of the transfer function with respect to its input, y_i is the current output of processing element i, and s is the training sample presented to the network. The output from linear programming model; inflow, storage, release and actual demand are given as input into ANN model of the basin. Monthly values of inflow, initial storage, demand and time period are the input into a three-layer neural network and output from this network are monthly release. The training set consisted of data from 1969 to 1994. The same data were used to test model performance as learning progressed. This study examines the effect of sample size and network architecture on the accuracy of neural network estimates for the known posterior probabilities. Neural network toolbox in MATLAB 6.1[8] Release 12, software is used to solve the developed model with four input variables and one output variables. The length of data for both input and output values is from 1 to 300. A better prediction is given by the three layers ANN model with 25 neurons in each hidden layer. This network is used to compute water release from Vaigai reservoir. It is also found that BPLM is quicker in the search for a good result.

Third, the special type of neural network called recurrent neural network (RNN) shown in Fig. 3 is also used to fit the flow data for reservoir operation problem. The release produced by RNN is compared with that of other neural network models for reservoir operation. The results are compared in terms of meeting the demand of the basin. The results are shown in Fig. 4. Comparison of performance by different methods of NN is exhibited in the plot.

Validation of the model: Once the training (optimization of weights) phase has been completed the performance of the trained network needs to be validated on an independent data set using the criteria chosen. The data of water year 1994–1997 are used for validating the developed ANN

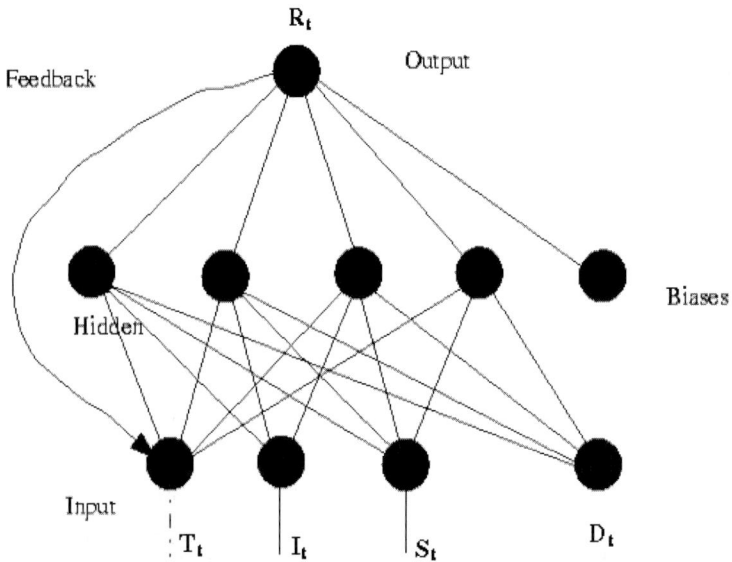

Fig. 3. Architecture of recurrent neural network.

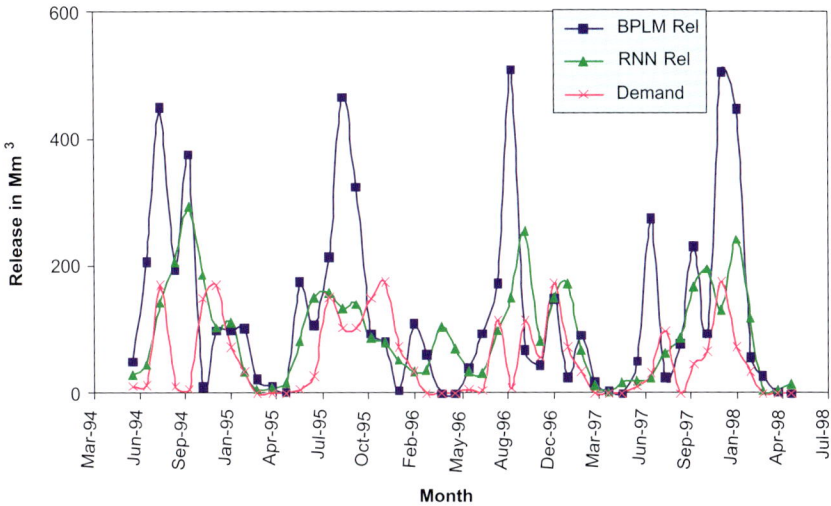

Fig. 4.   Comparison of performance by different methods of ANN.

model. The validation results are shown in Fig. 4. It is important to note that the validation data should not have been used as part of the training process in any capacity. The results are discussed in the following section.

4. Results and Discussion

The applicability of different kinds of neural networks for the probabilistic analysis modeling for random variables are summarized. The comparison comprehends two network algorithms (multilayer neural networks). This is a relevant result for neural networks learning because most of the practical applications employ the neural learning heuristic back-propagation, which uses different activation functions. Back-Propagation using Levenberg–Marguardt algorithm needs no additional executable program modules in the source code. But RNN has taken less epoch number compared to that of ANN. In addition, the number of epochs to find the optimal solution at different tests is significantly reducing in RNN. Further, the figures showed a relatively small running time taken to find the optimal solution when an RNN replaces the ANN model. This is exhibited in Fig. 4. The release from ANN, RNN and Actual practice are plotted for testing the performance of classical linear programming and heuristic method as applied to the problem of reservoir operation.

In the RNN model, the training parameters such as learning rate, momentum factor are progressively getting changed according to error obtained by the network. The number of epochs is kept very large so that the network will not terminate due to insufficient number of epochs. Thus only parameter to be changed is the number of recurrent neuron. The decrease of SSE for RNN models is found higher in all runs when compared with that of FFBPN models. It is found that the number of epochs is reducing with decrease in number of neurons in hidden layer. In all the run, the number of epochs required to train the RNN model are less compared to FFBPN. In the operation model for the Vaigai basin, RNN models have shown performance satisfactorily during validation phases. Also the data set resulted in a better representation when an RNN model replaces the ANN model using linear programming.

5. Conclusions

A study on intra neuronal, feedback network (RNN) proves to be a powerful operation tool that is drawn on the most recent developments in artificial intelligence research with a capability of not requiring assumptions about the underlying population. In this paper, the neural network approach for deriving operating policy for Vaigai reservoir is investigated. Overall results showed that the use of neural network model in operation of reservoir systems is appropriate. This developed model can be implemented for future operation of Vaigai system with the significance of no apriori optimization for getting release from the dam needs to be tested. A time series modeling on operation using LP with RNN provides a promising alternative and leads to better predictive performance than classical optimization techniques such as linear programming. The training time for intraneuronal network has been dramatically reduced comparing with that of BP networks (interneuronal) discussed here. Further work can be extended to three-dimensional information processing rather than from planar.

References

1. J. Sima, Back-propagation is not efficient, *Neural Networks* **9**, 6 (1996) 1017–1023.
2. S. Jiri, Backpropagation is not efficient, *Journal of Neural Networks* **9**, 6 (1995) 1017–1023.
3. T. A. Clair and J. M. Ehrman, Variations in discharge and dissolved organic carbon and nitrogen export from terrestrial basins with changes in climate: A neural network approach, *Limnol. Oceanogr* **41**, 5 (1996) 921–927.

4. C. C. Yang, L. C. Chang and C. S. Chen, Comparison of integrated artificial neural network with time series modeling for flood forecast, *Journal of Hydrosciences and Hydraulic Engineering* **17**, 2 (1999) 37–49.

5. V. Chandramouli and H. Raman, Multireservoir modeling with dynamic programming and neural networks, *ASCE Journal of Water Resources Planning and Management* **127**, 2 (2001) 89–98.

6. S. Mohan and V. R. Bai, Reliability based conjunctive use optimization model for crop planning, *Indian Journal of Power and Valley Development* **53**, 12 (2003) 215–221.

7. R. Lippman, An introduction to computing with neural nets, *IEEE ASSP Mag.* **4** (1987) 4–22.

8. MATLAB, software, Ver. 6.1.0.450 Release 12.1, Neural network toolbox for use with Matlab. User's Guide, 2001.

OPTIMIZATION MODEL FOR GROUNDWATER DEVELOPMENT IN COASTAL AREAS

NAMSIK PARK* and SUNG-HUN HONG

Department of Civil and Ocean Engineering, Dong-A University
Hadan-Dong, Saha-Gu, 840, Busan 604-714, Korea
*nspark@dau.ac.kr

An optimization model is developed for extraction and management of groundwater in coastal areas. The model consists of a sharp-interface flow model and a genetic algorithm optimization program. The flow model can simulate both freshwater flow and saltwater flow. An objective function is formulated so that the optimization model can determine not only the optimal extraction rates and locations but also the optimal control of an intruding saltwater wedge. For the latter, either freshwater injection or saltwater extraction can be considered. Impacts on groundwater environment are also considered. The proposed optimization model provides a versatile and comprehensive tool for developing and managing coastal groundwater.

1. Introduction

Pumping of groundwater affects both groundwater and surface water systems by interfering with the natural hydrologic cycle. Bear[1] indicated that extensive extraction of groundwater has upset the long established balance between freshwater and saltwater flows, causing the encroachment of saltwater into freshwater aquifers. As a large proportion (about 70%) of the world's population dwells in coastal zones, determination of the optimal development schemes of fresh groundwater and of the optimal control methods of saltwater intrusion is a big challenge for both engineers and managers to satisfy present-day and future water demands.

Recently, a number of studies have reported on optimal groundwater development in coastal regions.[2] In most of these studies, the decision variables are limited to pumping rates at predetermined well locations. Park and Aral[3] developed a model that can determine not only pumping rates, but also locations. However, a major shortcoming of previous studies stems from the use of analytical solutions. Use of analytical solutions limits the application of the model to very simple problems, namely, homogeneous and infinite aquifers, and stagnant saltwater.

A new optimization model is proposed in this study. The model can identify optimal rates and locations not only for freshwater pumping wells, but also for freshwater injection and saltwater pumping wells. The model can be applied to heterogeneous aquifers of finite size and irregular shape. Positive and negative impacts on groundwater environments are also considered. The optimization model proposed herein is an addition to the literature in that removes a significant portion of the limitations of previous models and extends the applicability.

2. Optimization Model

2.1. *Objective function*

The optimization process may become simpler by combining objective functions into a single function. The constraints also need to be combined into the objective function so that an unconstrained optimization technique, such as the genetic algorithm, can be used. The final form of the objective function in coastal aquifers becomes

$$\text{maximize} \quad \Phi = \alpha \sum_{i=1}^{N_{\text{opt}}} Q_f - \omega_1 \sum_{i=1}^{N_p} \frac{Q_s}{Q_f + Q_s} - \omega_2 \sum_{i=1}^{N_s} Q_s, \tag{1}$$

where Q_f and Q_s are the freshwater and saltwater pumping rates in the well i, N_{opt} is the number of targeted optimal wells, N_p is the total number of wells including both optimal wells and existing wells in a domain, N_s is the number of optimal wells designed to withdraw saltwater, ω_1 and ω_2 are the weighting factors. The first term can be considered a benefit, the second term is a penalty function that reduces the value of the objective function when saltwater is pumped from the wells designed to pump freshwater only, and the third term is the cost incurred from pumping saltwater to control the saltwater wedge. The number of wells is assumed to be determined *a priori*. α is a groundwater protection index representing the adverse impacts caused by pumping on the groundwater environment (Fig. 1). The groundwater protection index is defined as below:

$$\alpha = 1 - (D + I + A) \times \omega_3, \tag{2}$$

where, ω_3 is a weighting factor, D is the weighted normalized average drawdown ($= V_D/A_T H$), I is the weighted normalized average increase in saltwater thickness ($= V_S/A_T H$), A is the weighted normalized saltwater volume ($= A_I/A_T$), A_T is the domain area, V_D is the volume of drawdown, H is

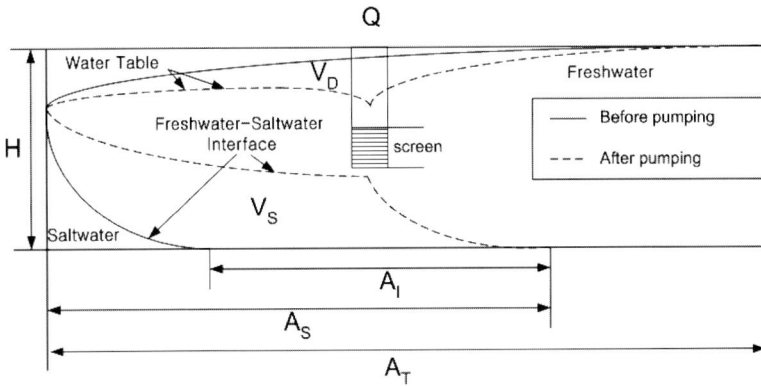

Fig. 1.   Adverse impacts of pumping in coastal areas.

the saturated thickness, V_S is the change in saltwater volume, and A_I is the change in intrusion area.

2.2. *Simulation model*

DUSWIM, which is used in this study, is a sharp-interface (two-phase) and quasi three-dimensional model. The mathematical model used in this design is based on two vertically integrated governing equations for the freshwater flow and the saltwater flow.[4] The aquifers can be either single or layered. A variety of transient and steady-state boundary conditions can be employed. These include: prescribed head and flux conditions, areally distributed recharge, well pumping or injection, vertical leakages through confining layers, and other head-dependent fluxes (e.g. at coastal boundaries).

2.3. *Applicability*

Modeling is generally performed under given conditions to assess the influences of prespecified pumping locations and rates. However, determination of optimal well locations and pumping rates often requires numerous simulations. An optimization method is needed to facilitate efficient search. Most groundwater development problems can be classified as one of the types specified in Table 1.

When the location of wells to be optimized is known, this type of problems is considered the first type. Optimal pumping rates would vary whether

Table 1.   Types of common groundwater development problems.

	A priori determined variables	Decision variables	Optimization
Stage 1: Optimal arrangement of pumping wells			
CATIA	Location	Freshwater pumping rate	Minimize adverse impacts
CATIB	Pumping rate	Location	Maximize pumping rate
CAT2A	—	Location and freshwater pumping rate	Maximize pumping rate while minimizing adverse impacts
Stage 2: Control of saltwater wedge			
CAT2B	—	Locations and freshwater pumping rate or injecting rate	Maximize pumping rate while minimizing adverse impacts
CAT4	—	Locations and freshwater pumping rate or saltwater pumping rate	Maximize pumping rate while minimizing adverse impacts

there are existing wells that may be affected by the new pumping wells or impacts on the groundwater environment are to be considered. When the desired pumping rates are predetermined, the problem reduces to finding the optimal locations. This type of problem is referred as the second type. Multiple optimal solutions may arise if impacts on the groundwater environment are not considered. The third type involves problems of finding both the optimal pumping rates and the optimal locations. When more groundwater pumping is required than the optimal pumping rate, a saltwater-wedge control method can be used. Freshwater injection and saltwater pumping can achieve the goal.

3. Example Application

The proposed optimization model is applied to an example problem. The modeling domain is an unconfined aquifer, 40 m thick, and 1.5 km × 1.5 km wide (Fig. 2).

The aquifer is assumed to be heterogeneous with four distinct permeability zones of four orders of magnitude differences. There is a coastal line on the right side. Freshwater enters the domain at the rate of $0.52\,\mathrm{m^3/day/m^2}$ through the left side. The densities of freshwater and seawater are 1,000 and $1{,}025\,\mathrm{kg/m^3}$, respectively.

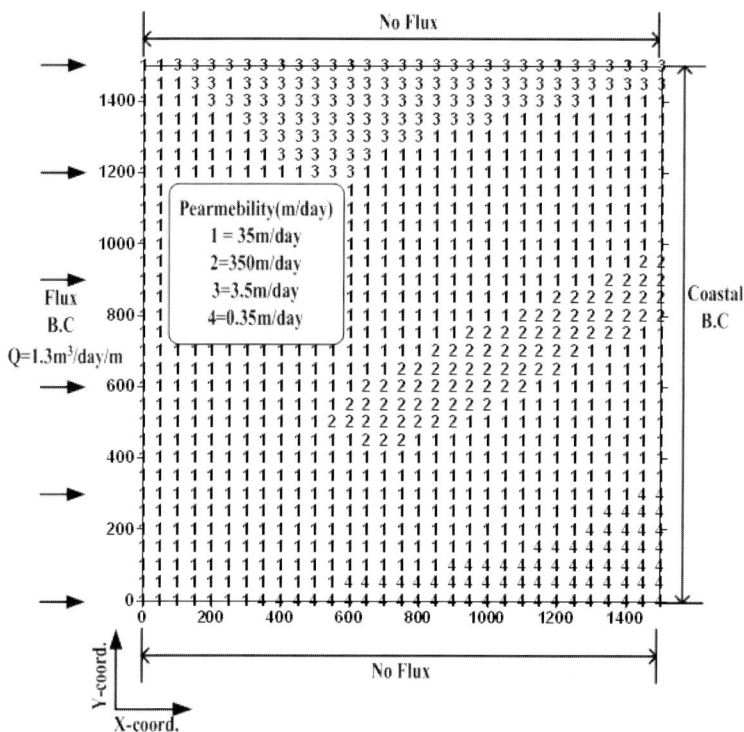

Fig. 2. The schematic diagram of unconfined aquifer for application.

3.1. *Optimal pumping rates*

Eleven new wells are planned at the predetermined locations (Fig. 3). All wells are screened down to 10 m below the mean sea level. Optimal pumping rates are calculated using the proposed model (Fig. 4). Total freshwater pumping rate of all the new wells is about $1,500\,\mathrm{m^3/day}$, and the maximum pumping rate is $460\,\mathrm{m^3/day}$ at well 4.

3.2. *Excessive pumping*

To demonstrate indirectly the optimality of the solution the pumping rate of well 3 is increased by $86.4\,\mathrm{m^3/day}$ higher than the determined optimal pumping rate to see if saltwater contamination occurs. Figure 5 depicts the ratio of the saltwater pumping rate and the total pumping rate along with the distribution of drawdown in freshwater hydraulic head. Figure 6 shows

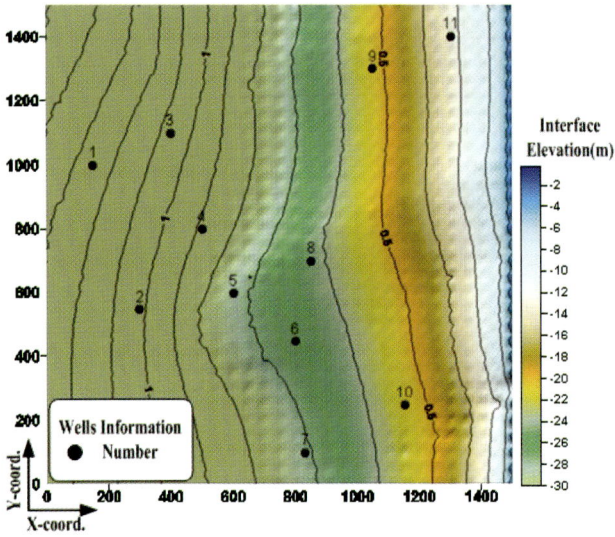

Fig. 3.   Well locations, water level and interface elevation in predevelopment condition.

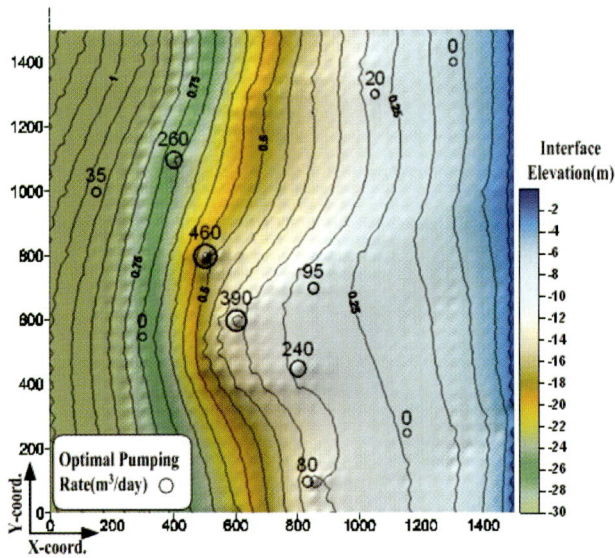

Fig. 4.   Optimal pumping rate and distribution of water level and interface elevation.

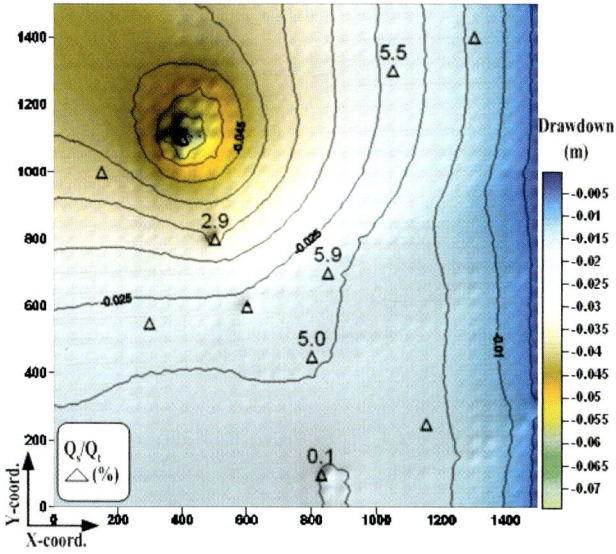

Fig. 5. Ratio of saltwater pumping rate and total pumping rate, and distribution of drawdown.

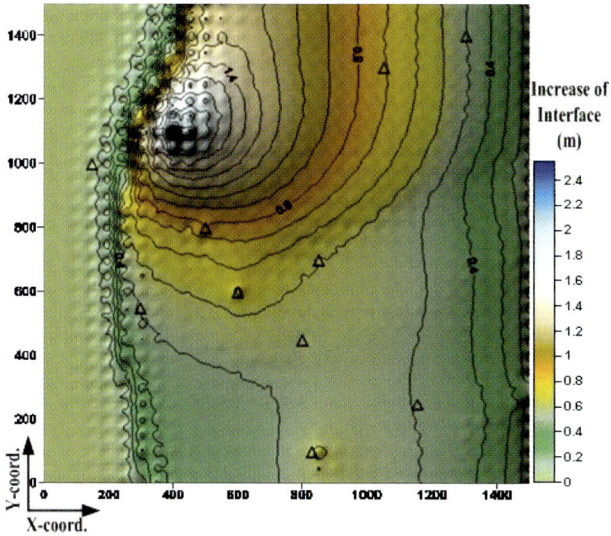

Fig. 6. Change of interface elevation caused by excessive pumping rate.

the increase of interface elevation caused by the excessive pumping from
well 3.

4. Results and Conclusions

A simulation-optimization model for development and management of
groundwater in coastal areas is developed. The proposed model can be
used to evaluate not only the optimal freshwater pumping rates and the
well locations but also the freshwater injection or the saltwater pumping
rates to protect freshwater pumping wells from saltwater intrusion. Integrity
of the groundwater environment is represented by three parameters indicating changes in the freshwater head, the saltwater intrusion area and
the saltwater volume. The use of a numerical sharp interface eliminates, at
the expense of more computer time, most of the limitations of analytical
solutions that are used in previous optimization models.

Acknowledgments

This research was supported by a grant (3-3-2) from Sustainable Water
Resources Research Center of 21st Century Frontier Research Program.

References

1. J. Bear, A. H.-D. Cheng, S. Sorek, D. Ouazar and I. Herrera (eds.), *Seawater
   Intrusion in Coastal Aquifers — Concepts, Methods and Practices* (Kluwer
   Academic Publishers, Netherlands, 1999).
2. S.-H. Hong, S.-H. Song, S.-K. Bae and N. Park, Verification and validation of an optimization model for groundwater development in coastal areas,
   Proceeding of 18th SWIM, Cartagena, Spain (2004), pp. 77–90.
3. C.-H. Park and M. M. Aral, Multi-objective optimization of pumping rates and
   well placement in coastal aquifers, *Journal of Hydrology* **290** (2004) 80–99.
4. N. S. Park, S. H. Hong, M. G. Shim, S. Y. Han and S. K. Bae, Optimization of groundwater withdrawal in coastal regions, *Proceeding of SWICA-M3*,
   Included in CD, Merida, Mexico, 2003.

SENSITIVITY ANALYSIS FOR OPTIMIZATION MODEL FOR COASTAL GROUNDWATER

SUNG-HUN HONG* and NAMSIK PARK

Institute of Construction Technology and Planning, Dong-A University
P.O. Box 604-714, Hadan-Dong, Saha-Gu, 840, Busan, Korea
wghsh72@smail.donga.ac.kr

A simulation-optimization model can be a useful tool in decision-making regarding groundwater developments. A flexible model is developed to determine optimal solutions of not only the distribution of the pumping rates and the locations, but also control measures to prevent or reduce saltwater intrusion. This paper discusses the result of a sensitivity analysis for the model. Influence of hydraulic conductivity, recharge rate and aquifer thickness on optimal pumping rates and optimal well locations is studied. It is shown that the hydraulic conductivity has more pronounced impact than other parameters on optimal pumping rates and optimal well locations.

1. Introduction

Sensitivity analysis is important in understanding not only the characteristics of the model, but also the behavior and influence of various parameters in the model. The general purpose of a sensitivity analysis is to quantify the uncertainty in the calibrated model caused by uncertainty in the estimates of aquifer parameters, stresses, and boundary conditions.[1] Tran[2] studied the response in saltwater intrusion lengths with respect to stresses and transmissivities using a SHARP simulation model. The objective of this study is to observe how the optimal solutions change depending on the parameters. Sensitivity analysis of major parameters on optimal solutions is investigated to aid potential users of the model.

2. Sensitivity Analysis for Optimization Model

A hypothetical cross-sectional unconfined aquifer (Fig. 1) is used to conduct the sensitivity analysis. The hydraulic conductivity (K), the total recharge rate (Q_{rech}), and the aquifer thickness (B) are selected as control parameters. The optimal pumping rate (Q_{opt}) and the optimal location of the well (L_{opt}) are state variables that are affected by the control parameters.

S.-H. Hong and N. Park

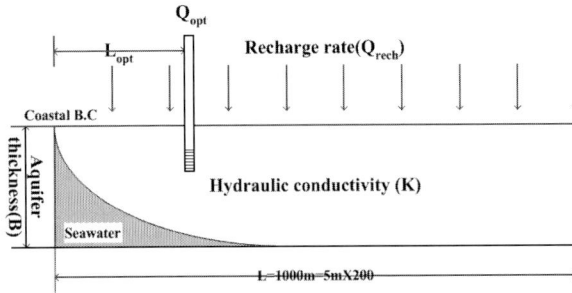

Fig. 1.   The schematic diagram of unconfined aquifer for the sensitivity analysis.

Table 1.   Investigated major parameters.

Area	Depth of alluvium (m)		Hydraulic conductivity (m/sec)		SGD/precipitation (%)	
	Min.	Max.	Min.	Max.	Min.	Max.
Nakdong River	5.2	15.2	15.2	344.23	2.93	3.2
Youngsan Somejin River	6	20	30.77	248.48	2.11	2.48
Keum River	5.7	9.9	10.77	24.56	0.57	0.64
Buan-Gun	3.6	30	16	28	—	—

Table 1 presents the ranges of the hydraulic conductivity, of the recharge rate and of the aquifer thickness deduced from previous studies.[3-6] Three problems types are considered: wells with fixed locations, wells with fixed pumping rates, and wells with both locations and rates are to be determined. The average annual precipitation (1,283 mm/year) of Korea is used to calculate the recharge rate.

2.1. *Sensitivity analysis for the optimal pumping rate*

In this section the sensitivity of the optimal pumping rate is investigated. Figure 2 shows the perturbation of the optimal pumping rate as a function of the change in the recharge rate and the hydraulic conductivity. For convenience the pumping rate is normalized by the total recharge rate. As is seen from the figure, the dimensionless optimal pumping rates are small in highly conductive aquifers and in low recharge regions. The optimal pumping rate increases with curvilinear relationships according to the recharge rate in the low-conductive aquifer. However, the optimal pumping rate varies nearly linearly with the recharge rate in highly conductive aquifers. The optimal

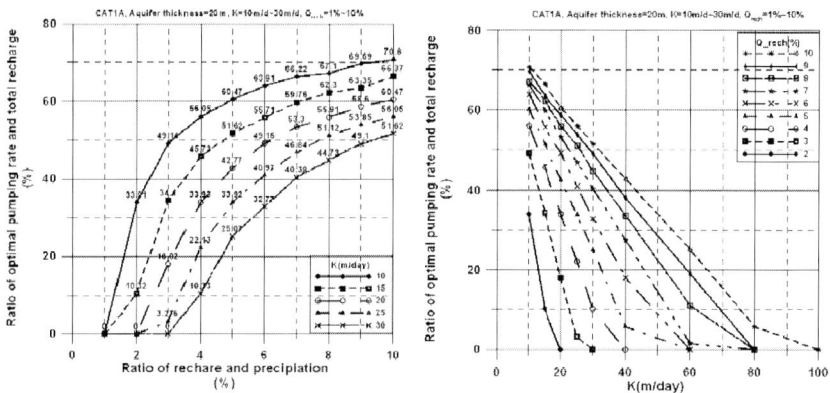

Fig. 2. Normalized optimal pumping rate versus recharge rate and hydraulic conductivity.

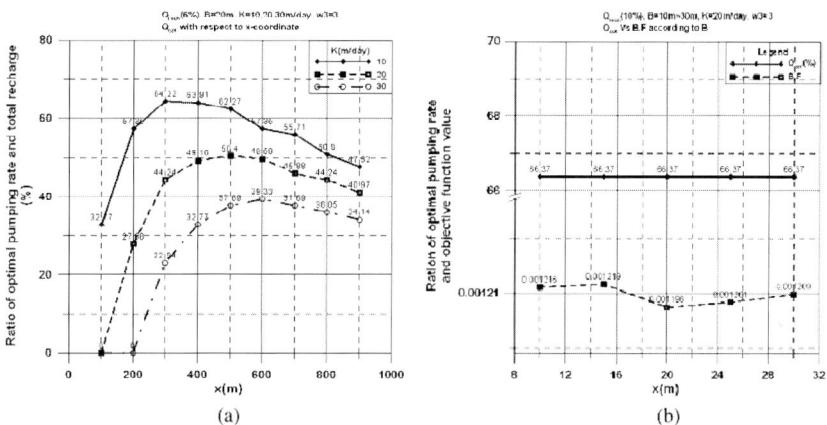

Fig. 3. (a) Optimal pumping rate versus hydraulic conductivity via enumeration, (b) optimal pumping rate versus objective function value with various aquifer thicknesses.

pumping rate decreases quickly in low recharge regions. Meanwhile, it has linear relationship with the hydraulic conductivity in high recharge rate regions.

The relationship between the optimal pumping rate and the hydraulic conductivity is investigated via enumeration for the specified recharge rate (Fig. 3(a)). As is seen from this figure, the optimal pumping rate is higher in low conductive aquifers than in highly conductive aquifers. However, the

range of optimal pumping rates is more sensitive to the well location in
the low-conductive aquifer. The rate varies nearly 20% in low-conductive
aquifers whereas it varies only about 5% in highly conductive aquifers.

Figure 3(b) shows the relationship between the optimal pumping rate
and the objective function value as a function of the aquifer thickness. Even
though aquifer thickness does not affect the optimal pumping rate, it does
affect the objective function value. Therefore, the optimal pumping rate is
independent of the aquifer thickness.

2.2. Sensitivity analysis for the optimal well location

In this section, sensitivity of the optimal well location with respect to the
pumping rate is discussed. Figure 4(a) shows the optimal well locations
for the given pumping rates with various hydraulic conductivities. With
the same increase of the pumping rate from 10% to 40%, the optimal well
locations move approximately 50, 100, and 280 m depending on the increase
of hydraulic conductivity. The optimal well location in highly conductive
aquifers is less sensitive than the optimal well location in low-conductive
aquifers.

Figure 4(b) presents the change of optimal well location depending on
both the hydraulic conductivity and the recharge rate. The pumping rate
is specified as 50% of the recharge rate. The optimal well location is less

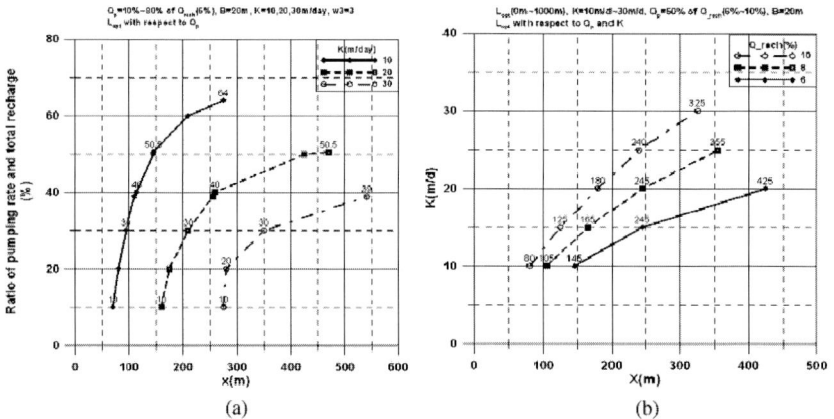

Fig. 4.   (a) Optimal well location versus the pumping rate with various hydraulic conductivities and, (b) optimal well location versus hydraulic conductivity with varying
recharge rates.

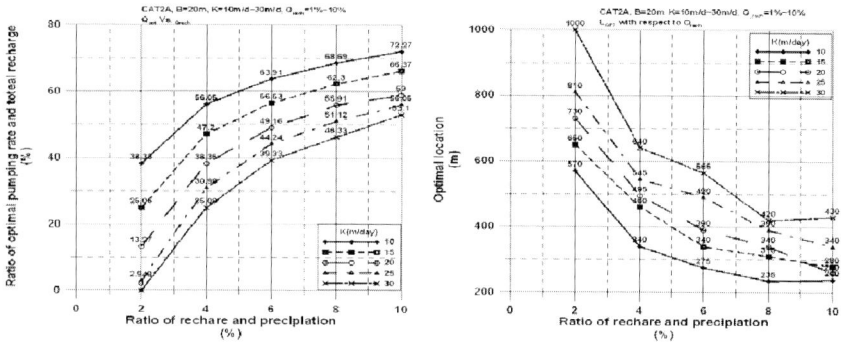

Fig. 5.   Optimal pumping rates and optimal well locations versus the recharge rate with various hydraulic conductivities.

sensitive in high recharge rate regions than low recharge rate regions and moves toward coastline according to the increase of the recharge rate.

2.3. *Sensitivity analysis for the optimal pumping rate and well location*

Sensitivities of both the optimal pumping rate and the optimal well locations are investigated with respect to changes in the recharge rates with various hydraulic conductivities (Fig. 5). The shape of the optimal pumping rate curve is similar to that in Fig. 4. However, the optimal pumping rate is higher than the results presented in Fig. 2. That is to say, simultaneous consideration of the optimal pumping rate and the optimal well location can yield more groundwater.

3. Results and Conclusions

The optimal pumping rate decreases with respect to the increases in hydraulic conductivity. Meanwhile, the relationship between the optimal pumping rate and the recharge rate differs depending on the hydraulic conductivity. The relationship is curvilinear in low-conductive aquifers, but becomes linear in highly conductive aquifers. The optimal pumping rate is more sensitive to hydraulic conductivity in low recharge regions. The aquifer thickness does not affect the optimal pumping rates. Optimal well locations change quickly inland depending on the increase of hydraulic conductivity and are more sensitive to hydraulic conductivities than recharge rates.

In conclusion, we found that accuracy in estimated hydraulic conductivity is more important in determining optimal pumping rates and well locations than accuracies in recharge rates and aquifer thicknesses. Contrary to inland groundwater developments where saltwater intrusion is not a concern, we found that in coastal areas more groundwater pumping is possible in low-conductive aquifers than in high-conductive aquifers.

Acknowledgments

This research was supported by a grant (3-3-2) from Sustainable Water Resources Research Center of 21st Century Frontier Research Program.

References

1. M. P. Anderson and W. W. Woessner, *Applied Groundwater Modeling Simulation of Flow and Advective Transport* (Academic Press, San Diego, 1992).
2. T. M. Tran, *Multi-Objective Management of Saltwater Intrusion in Groundwater: Optimization under Uncertainty* (DUP Science, The Netherlands, 2004).
3. Korean Water Resources Corporation (KOWACO), Regional groundwater investigation report on Young-San River and Seom-Jin Rive Basin, KOWACO, 1998.
4. KOWACO, Regional groundwater investigation report on Nak-Dong River Basin, KOWACO, 2000.
5. KOWACO, Regional groundwater investigation report on Kum River Basin, KOWACO, 2002.
6. Korea Agricultural and Rural Infrastructure Corporation (KARICO), Groundwater management investigation report on Budong Area of Buan-Gun, KARICO, 2003.

MODELING THE GROUNDWATER DYNAMICS IN A SEMI-ARID HARD ROCK AQUIFER INFLUENCED BY BOUNDARY FLUXES, SPATIAL AND TEMPORAL VARIABILITY IN PUMPING/RECHARGE

M. SEKHAR[*,§], S. N. RASMI[*], Y. JAVEED[*], D. GOWRISANKAR[†] and L. RUIZ[‡]

*Department of Civil Engineering, Indian Institute of Science
Bangalore 560 012, India
†RRSSC-CMO, Department of Space, ISRO, Bangalore 560 070, India
‡INRA, Renne, France Visting Scientist, IFCWS, Indian Institute of Science
Bangalore, India
§muddu@civil.iisc.ernet.in

Regional groundwater modeling is important for assessing the groundwater balance and proper management of groundwater system especially in semi-arid regions of hard rock aquifers. This study deals with groundwater modeling of the Gundal subbasin, which is located in the semi-arid portion of the Cauvery river basin (India). The interesting feature of this subbasin is the dramatic declines of groundwater levels in some areas close to recharge zones due to the impacts of land use and land cover changes. An integrated groundwater modeling approach is adopted for a better assessment of water balance components. The modeling is supported through the "soft" data obtained from remote sensed platforms in addition to the conventional hydrogelogical "hard" data. The modeling study shows the impact of pumping resulting in regional groundwater flows influencing the hydrogeologic regime in the recharge zone of the subbasin, which is located in the forested region of the Bandipur national park. The hypothesis of interbasin transfer of subsurface flow from the neighboring Nugu river basin and the relevance of appropriate boundary condition for the subbasin is discussed based on the water balance computations.

1. Introduction

In the semi-arid regions of the Cauvery river basin, the changing land use to intensely developed agriculture is highly dependent on irrigation with surface water and groundwater to meet crop water demands over the Kharif and Rabi seasons. The agriculture water use and groundwater storage changes observed suggest a need for groundwater management in this basin similar to the concerns voiced in the semi-arid basins elsewhere in the world.[1] Numerical modeling is routinely employed for analyzing the problems associated with groundwater flow at watershed, subbasin and basin

173

scales. Most of the groundwater models are distributed models and parameters used are not directly measurable. Recharge estimation using models based on groundwater level fluctuation method have been discussed in several studies.[2] Estimation of groundwater pumping, which is frequently the least measured water balance component in semi-arid basin with significant agricultural production is discussed using GIS-based water balance model.[3]

The present study attempts to model a regional groundwater system to analyze groundwater flow in a hard rock aquifer in the Gundal subbasin of the Kabini river basin, which is very important for groundwater assessment and management in this area. In this application groundwater modeling is combined with remote sensing and GIS approaches for parameterization of the subbasin. A two-dimensional fully distributed groundwater model on the concept of predominantly lateral flow conceptualized as an unconfined aquifer has been used in this study for calibration. The model is applied to study the spatial and temporal variability in the groundwater pumping due to changing land use practices over a period of 25 years affecting the spatial and temporal flux distributions within various zones of the subbasin and the associated controls on the boundary conditions with the neighboring subbasins.

2. Setting

The Gundal subbasin is located in the south west of Karnataka state. The location map of study area is shown in Fig. 1. The Gundal subbasin occupies an area of 1,270 km^2 and it stretches from $76°30' - 76°51'\ 47''$ longitude and $11°40'13'' - 12°7'13''$ latitude. The study area mainly consists of one major rock type, which is granitic gneiss. It is observed that the lineaments in this area are drainage oriented and fracture controlled. Many of the lineaments in this basin are rectilinear type and trend north-south or NNE-SSW and range from 2.5 to 7.5 km in length.[4] The vegetation in Gundal watershed is characterized by agriculture activity. Traditionally, crops are grown during kharif (monsoon) and rabi (dry) seasons. Main traditional crops are ragi (finger millet) and pulses, whereas paddy is grown in the command areas of tanks and canal command areas (northern part of the basin which forms the discharge area). Since the last two decades, the major source of water for irrigation in the rest of the subbasin is groundwater, allowing double-crop cultivation. As a result of increased irrigation by bore wells, irrigated crops like sugarcane and cash crops replace traditional rainfed crops.

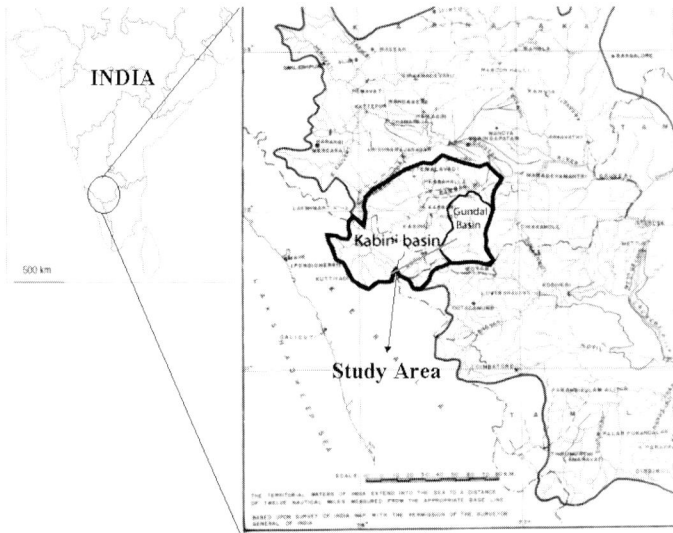

Fig. 1.   Location map of Gundal subbasin in Cauvery river basin.

The thematic layers of land use and structure are prepared from the remote sensing and are shown in Fig. 2. It may be observed from the two-season land use map (Fig. 2(a)) that substantial double crop areas (kharif + rabi) exist not only in the discharge zone but also in the recharge areas of the subbasin. The double crop areas in the recharge zone are managed by groundwater pumping. It may be noted that sustained pumping for both kharif and rabi crops in these zones are possible due to the good yields in these areas, which are correlated to the presence of several lineaments and structural control of groundwater in these parts. Figure 2(b) shows the structural map of the subbasin, which shows the reasons for the sustainability of double crop in some of the recharge and intermediate zones of the subbasin. These thematic layers along with the hydrogeomorphology, soil and transmissivity (obtained from pump tests) maps are combined to produce an integrated composite zonation map (Fig. 2(c)), which delineates nine homogeneous zones. Each of the zones contains one observation well (OW) which is referred with the corresponding zone number (i.e. OW7 refers to well in zone 7). Figure 2(c) shows also the locations of the monitoring network stations for the rainfall and groundwater level data used for carrying out the analyses in the subbasin during the period 1977–2000. The draft has been increasing in this subbasin beginning 1975. A detailed

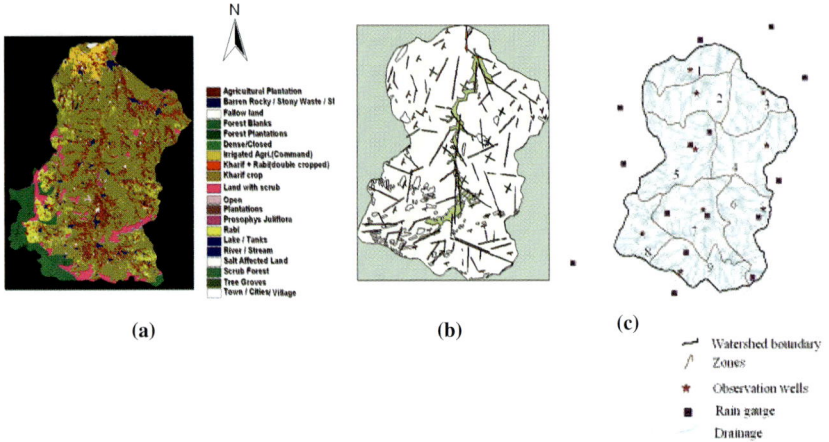

Fig. 2.   Thematic maps of (a) land use, (b) structures, and (c) map showing rain gauge, observation well network along with zonation structure for the subbasin.

draft data is obtained for each of the villages of the subbasin during 1992–1994, which is used to assess draft for each of the subzones. These data, along with the temporal variation of draft estimated at taluk (administrative unit) is used to estimate the draft for each of the years from 1979 to 2000 for all the subzones.

3. Results and Discussion

3.1. *Groundwater dynamics*

The distributed groundwater flow model Visual MODFLOW is used for simulating the groundwater behavior in the subbasin, combining the inputs from the field data collected for the homogeneous zones. It is assumed that the study region can be described as an unconfined groundwater aquifer at a regional scale. A simple rainfall recharge relationship is used, assuming recharge as a linear function of rainfall. For the recharge component from return flows due to surface and groundwater irrigation areas, recharge is assumed to be a percentage of the amount of irrigation depending on the type of crop and type of irrigation, based on the norms available for this region.[5] MODFLOW is calibrated during the period from May 1977 to April 1979 for the entire subbasin comprising of nine homogeneous zones. The calibration period pertains to a uniform draft scenario in most of the subbasin, in addition to a considerably lower draft in comparison to later years. The calibration is performed assuming specified transmissivities for each of the zones obtained from several pump tests in the region[6] along with an assumption of no flux boundary along the entire subbasin. In this case, the calibration provides an estimate of recharge factor and specific yield for each of the zones, along with the lateral groundwater fluxes between the zones. The fit between the observed and simulated groundwater levels during the calibration period is found to be good (Fig. 3). Before applying MODFLOW for the subbasin a lumped ground water balance model is considered for zone 7. In this model, the lateral inflow and outflow are maintained constant throughout the entire simulation period. The model is calibrated with suitable inflows and outflows, and recharge for the period between 1977 and 1979. The system is simulated using either a constant draft (based on the draft of 1979) or a variable draft for the entire period up to year 2000. Figure 4 shows the simulations for these cases and their comparison with the observed water levels in OW7. It is clear from the plot that though the lumped model provides a good fit up to 1984, thereafter both the constant and variable draft cases fail to match the observed water level responses. The constant draft produces a response as if the groundwater levels are quite stable over a long period from 1984 to 2000. On the other hand, the variable draft produces a monotonically decreasing groundwater levels as expected. The simulated water levels for the variable draft case show a trend similar to observed pattern up to 1990 with a marginally lesser decline. However, the simulation is not able to reproduce the observed raise and subsequent fall in the later periods. Thus, these variations might

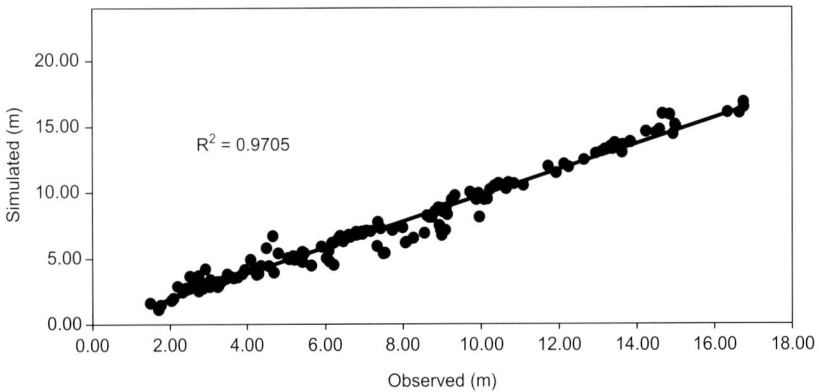

Fig. 3.   Model performance for the subbasin during calibration period (1977–1979).

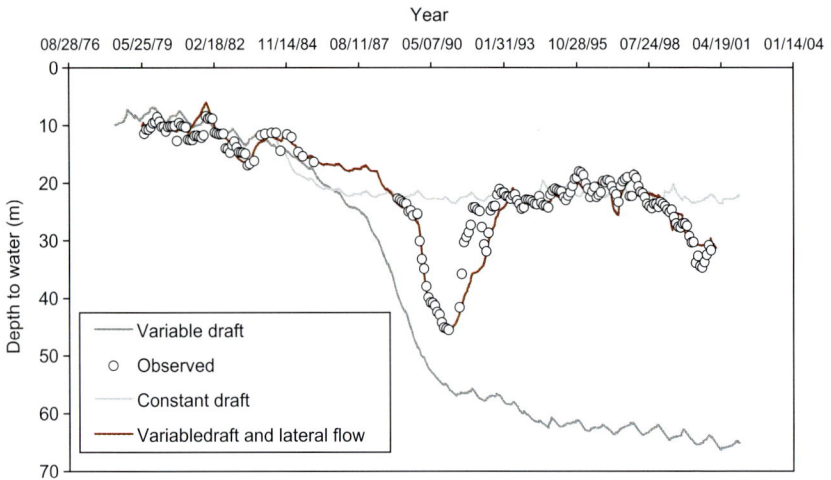

Fig. 4.   Groundwater level simulations obtained at OW7 for various cases.

be due to the impact of later inflows into zone 7, which are determined by
the upstream zones of 8 and 9. This analysis suggests the need for considering groundwater modeling of the entire subbasin connecting all the zones
rather than a lumped model for individual zones alone.

3.2. Modeling the groundwater fluxes

The calibrated model is used to simulate the behavior of the groundwater
system from 1979 to 2000 using a variable draft pattern but with a no flux
boundary condition for the subbasin. Figure 5 shows that the simulated
responses for the water levels in OW8 and OW9 is in good agreement with
observed data up to 1984, but not in the later periods. More importantly,
the simulated water levels show an increasingly declining trend with years in
contrast to the observed responses. This might be attributed due to higher
drafts in zone 7 causing higher groundwater fluxes from zones 8 and 9 to
zone 7. In order to improve the performance of the model simulations, the
assumption of the no flux along the boundary forming zones 8 and 9 needs to
be relaxed.[7] Simulations are performed again for calibrating the boundary
fluxes in each year after 1984, in such a way that a good match is obtained
between the simulated and observed water levels in OW9 and OW8 (Fig. 5).
This approach gives good match with the observations in the wells in the
other zones 1–7. Figure 4 shows the fit obtained for OW7. The inflows into

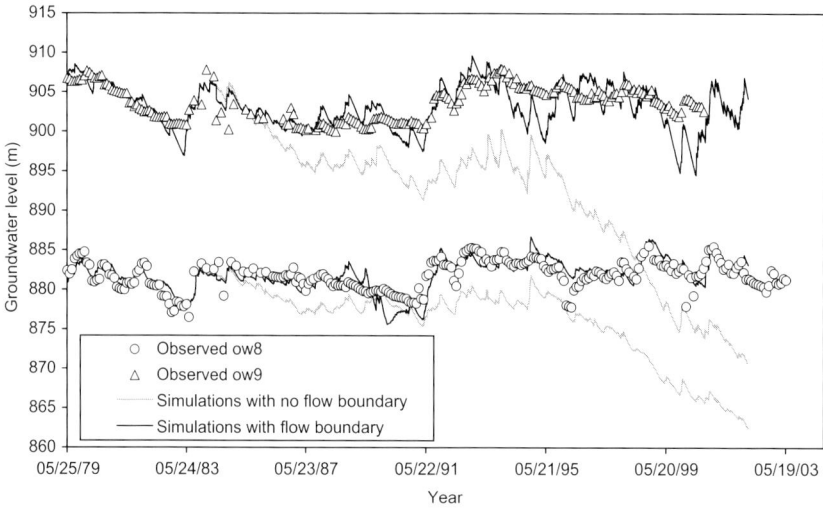

Fig. 5. Groundwater levels simulations obtained at OW8 and OW9 for various cases.

all these zones show an increasing trend. The inflow is the highest for the zone 7, which receives groundwater inflows from both zones 8 and 9. The inflows into zones 9 and 8 might be due to the groundwater fluxes from the neighboring Nugu subbasin adjoining the boundaries of zones 8 and 9 located in the Bandipur national park.

3.3. *Groundwater balance*

The groundwater balance for the high pumping zone of 7 is shown in Table 1 for the period 1979–2000. The recharge factor is approximately 3.5% in zone 7, which correlates with the different soil types found in this zone. The net lateral flow into zone 7 is approximately 70% of the draft in this zone for each of the periods. Thus, higher drafts over the years are sustained to some extent from the increasing inflows into this zone from adjacent zones of 8 and 9. The inflows into this zone have increased by 70% while draft

Table 1. Average annual groundwater balance for zone 7 for a season.

Season	Rainfall (mm)	Recharge (mm)	Inflow (mm)	Outflow (mm)	Draft (mm)	Storage change (mm)
1979–1984	673.71	23.58	79.86	21.37	82.67	−0.50
1984–1989	597.96	20.94	83.75	16.22	100.27	−11.82
1989–1994	695.30	24.33	102.47	13.55	105.85	7.42
1994–1999	804.36	28.15	110.10	12.98	129.47	−4.80

has increased by 63% over two decades (1979–1999). In spite of these large inflows, the storage change in this zone shows a significant decline during drought periods (1984–1989 and after 1999) due to lower recharge from rainfall in this zone as well as correspondingly reduced inflows from the adjacent zones. On the contrary during good rainfall periods (1991–1993) an increase in the groundwater storage is observed. In the zone 9, the draft is quite small and is approximately 30% of the recharge. The pumping is not much changed in this zone as considerable part of this zone lies in the forested region. The outflows from this recharge zone have increased by about 35% during 1979–1999. Further inflows into this zone have increased considerably by 60% during 1984–1999 to sustain the requirements of zone 7. These inflows might have occurred from the adjacent region of Bandipur national park located in the Nugu basin. Due to these inflows, the storage change in this zone shows an interesting trend of increasing water levels in spite of high pumping occurring in adjacent zone 7.

4. Conclusions

The Gundal subbasin, which is located in a semi-arid portion of the Cauvery river basin, is intensively cultivated through irrigated canal command in its Northern part and groundwater in the recharge and intermediate zones. Significant part of the recharge zones is part of the ecologically sensitive Bandipur National Sanctuary. It was found that the evaluation of the impact of groundwater based agriculture needs to be carried out at the scale of the whole system. The most salient feature of the study is the necessity of analyzing groundwater balance at a regional scale, due to the importance of lateral fluxes, which are generally neglected in hard rock aquifers. The lateral groundwater flow appears to be controlled by the geological structures in the vicinity of high pumping zones. As a consequence, the assessment of the sustainability of agricultural practices in the high pumping areas located in the recharge zones of the subbasin, must take into account the impact on the neighboring zones, including the protected forested areas. The study indicates that water levels in the excessive groundwater depletion zones are sustained at the present level due to the inflows from the adjacent Nugu river basin. This might result in decline of water levels in the region of Bandipur national park.

Acknowledgments

This work was carried out through the support received from Indian Institute of Science, the French Institute of Research for Development and

Indian Space Research Organization. We also thank the Department of Mines and Geology (Karnataka) and Central Ground Water Board for allowing us to access the database.

References

1. P. Gleick, The changing water paradigm: looking at twenty-first century water resources development, *Water International* **26**, 1 (2000) 127–138.
2. M. A. Sophocleous, Interaction between groundwater and surface water: the state of the science, *Hydrogeology Journal* **10** (2002) 52–67.
3. N. Ruud, T. Harter and A. Naugle, Estimation of groundwater pumping as a closure to the water balance of a semi-arid, irrigated agricultural basin, *Journal of Hydrology* (in press).
4. K. C. Subhaschandra and G. Narayanachar, Structural features and groundwater occurrence in Gundal river subbasin, Mysore district, Department of Mines and Geology, Karnataka, Bangalore, No. 345, 1997.
5. GEC, Ground water estimation committee report, Ministry of Water Resources, Government of India, 1998.
6. CGWB, Hydrogeological conditions in Chamarajanagar District, Karnataka. Report of Central Ground Water Board, Ministry of Water Resources, South Western Region, Bangalore, June 1999.
7. M. Sekhar, S. N. Rasmi, P. V. Sivapullaiah and L. Ruiz, Groundwater flow modeling of Gundal subbasin in Kabini river basin, India, *Asian Journal of Water, Environment and Pollution* **1**, 1–2 (2004) 65–77.

DELINEATION OF WATER BEARING FRACTURES IN BORE WELLS BY EC LOGS

V. K. SAXENA*, N. C. MONDAL and V. S. SINGH

National Geophysical Research Institute, Hyderabad 500 007, India
vks_9020010@yahoo.co.in

The mostly geological and geophysical techniques are in practice for the delineation of water bearing fractures in the hard rock areas. The attempt has been made to develop some technique, which may be simple, more informative and also cost effective as well as give groundwater quality information. Electrical conductivity measurements were conducted in a large number of shallow bore wells (experimental) in hard rock areas. Electrical conductivity logs were carried out in three different locations of India: (1) Maheswaram: nine shallow bore wells in a granitic aquifers (2) Wailpally: four shallow bore wells in a granitic aquifers and (3) Sadras: four shallow bore wells in a charnokite aquifers. The observations were made at short interval of about 1 m from water table (narrow spacing wherever required) till the bottom of the bore wells. It is observed that EC have shown remarkable changes, which suggests the probability of water bearing fractures in the bore wells. In addition, the results of EC logs were compared with the results of various known geological and geophysical techniques. The results of EC logs are in good agreement with the results obtained by litholog profiles, inflow measurements, λ-logs, resistivity logs etc. Thus EC log technique is considered to be a new technique for the delineation of water bearing fractures in hard rock areas.

1. Introduction

In the study of the hydrology of fractured rocks, knowledge of the fracture properties is essential. Surface observations may be useful, but the more relevant observations are those made at the depths of interest. Lee developed a technique, for locating the contaminated groundwater discharge by mapping, variations in EC of bottom sediment and used in shallow surface water in an artificially created seepage area of a small shallow lake.[1,2] EC of water usually uses to get broad picture of TDS of water.[3,4] EC is also use to test the suitability of water for different purposes such as drinking, irrigation and various types of industrial applications etc. Most such measurements are made through boreholes or underground openings. In the case of borehole, various methods of determining fracture properties have been used.[5-7] For example, a down hole televiewer can be used to map the fracture traces on the borehole wall and determine their density and orientations.[2,8] However, it is well known that not all of these traces will correspond to water conducting fractures. Since many of the fractured rocks of interest are of low permeability, the flow rate from a packed interval can be very low. Packed-off test intervals are usually larger than individual water-conducting zones, thus leading to uncertainty in the location of a water-bearing fracture. In general, the flowing fractures contain fluids with different chemical composition and ion content from each other and hence have different electric conductivities.[9] Like an experimental technique, EC logging is subject to detection limits. For inflowing formation water is determined by the ability to properly identify EC changes caused by the inflow.

Microprobe could successfully detect discharges from both the shallow, local flow system and the deeper, more regional, bedrock flow systems.[10] Revil and Leroy showed the groundwater flow is responsible for an electric field called the streaming potential, which is the main component of the so called self-potential anomalies.[11]

In the present context EC logs has been used for identifying the depth of weathered zone/water bearing fractured rocks in bore wells. Present work summaries the results from 17 observational bore wells in three different hard rock problematic areas, and findings are compared with other geophysical/geological outputs. The present study describes a technique involving the use of electrical conductivity logs of bore well water (without of packers) for identifying the water bearing fractures in hard rock areas. In addition to this, technique may provide valuable information about the water table, temperature of water at different depths, EC at different depths, quality of water and saturation index at aquifer level etc.

2. Background of the Areas

The study area compared with three different locations in India (Fig. 1). (a) Maheshwaram watershed is situated in the R.R. district of Andhra Pradesh, India, about 30 km south of Hyderabad, covering an area 60 km^2, is a representative watershed in granitic rocks. The depth of water level in this area varies from 14 to 26.25 m. Groundwater flow is mainly associated with fractured rock aquifers. (b) Wailpally watershed is located about 60 km towards east of Hyderabad, covering about 50 km^2, also represented granitic terrain. Groundwater occurs in shallow-weathered and deep-fractured granite and in weathered gneisses. The depth to water level varies from a few meters to about 26 m below ground level. The groundwater is mostly

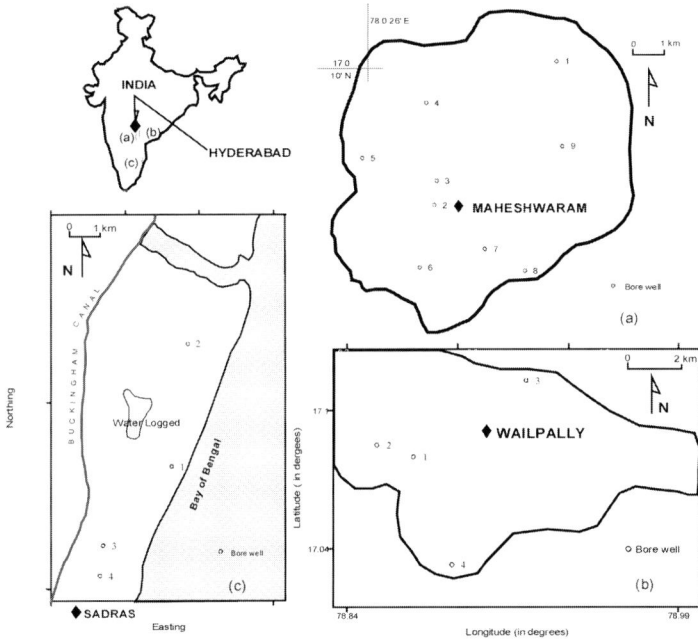

Fig. 1.   Location of experimental bore wells in three different watersheds, India.

exploited through dug wells, dug-cum-bore-wells and bore wells for irrigation, as well as domestic purposes and (c) Sadras watershed is also located
about 60 km southwest of Chennai and 10 km SS-W of Mahabalipuram close
to the Sadras. This watershed is spread in around 12 sq. km and located in
between Buckingham Canal and the coastal site of the Bay of Bengal. These
areas consist of hard crystalline rock mainly charnokite of Archaean age.
Groundwater level in the area fluctuates a few meters to about 20.5 m below
ground level. Groundwater velocities as determined by single well as well
as multi-well tracer measurements were in order of ≤ 4 cm/day.[12,13]

3. Instrumental Techniques and Procedure

Electrical conductivity logging of water with depth interval in 17 bore
wells have been carried out by using EC logger namely Wissenscnscn-
naftlich Technische-Werkstatten (WTW) of German make. This instrument
builds up of very sophisticated EC meter and logger with sensor with the
facilities of continuous EC measurements with depths upto 200 m along with

temperature monitoring device-recording system. Sensor attached with logger poured into bore wells and depths can be measures with its graduated cable. The observations were taken at 1 m interval. When sensor interact the water, EC meter record the electrical conductivity, temperature and depth. Laboratory study shows that even a minute mixing was preciously recorded.[14]

4. Results and Discussions

EC log profiles with depth have been carried out in nine bore wells during June 2001 (pre-monsoon) at Maseshwaram watershed, four bore wells in Wailpally watershed during October 2003 (post-monsoon) and also four bore wells in Sadras watershed in February 2004 (post-monsoon). The locations are shown in Fig. 1. One typical EC log profilings in Wailpally watershed is shown in Fig. 2, which indicating the location of fractured zones.

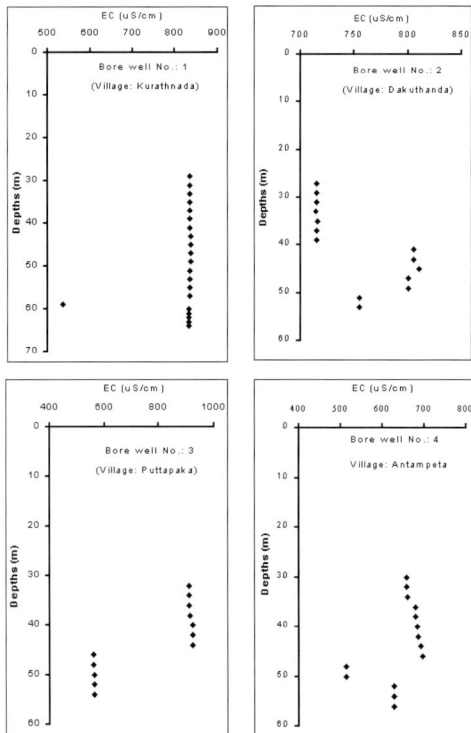

Fig. 2.   Depth vs. EC plots in Wailpally watershed, A.P., India.

Table 1. Depth of identified fractures zones in shallow bore wells by geophysical and geological information in Maheswaram watershed, AP, India.

Bore well number	Lithologic information	EC log	Water flow injection test	Gamma log	Resistivity log		
					LN	SN	PR
1	Fractured granite at 22.86–36.06	23.50–25.50 27.25–31.25 37.00–38.75	18.00–19.00 31.00–38.00	22.40–24.40 26.00–28.00 30.80–31.00	30.40–31 20	30.30–32.40	31.80–33.00
2	Semi-weathered granite at 30.14–36.00	36.00–37.50	—	—	34.70–37.00	34.00–35.30	36.10–38.50
3	Fractured granite at 23.00–32.20	19.70–21.25	—	34.40–34.80	18.70–22.50	28.80–37.00	27.70–29.80
4	Fractured granite at 18.00–18.50	19.70–21.25	—	28.00–29.00	20.10–22.00	26.70–29.00	27.50–29.00
5	Fractured granite at 29.87–33.53	25.70–29.20	29.00–30.00	18.00–23.00 27.90–33.20	28.30–34.50	26.80–32.40	28.40–34.40
6	Minor fractured granite at 15.84–32.90	24.00–25.70	18.25–24.25 26.00–27.50	26.00–26.80 34.00–34.80	30.50–36.00	29.00–34.50	31.00–35.00
7	Fractured granite at 17.68–23.70	17.20–19.50	17.50–18.50 18.50–21.00	—	20.20–24.50	18.75–19.20	18.10–26.80
8	Fractured granite at 19.20–23.78	15.75–20.00	18.00–22.00 28.00–29.00	18.20–24.60	14.00–34.00	14.00–34.00	20.70–26.80
9	Minor fractured granite at 13.80–23.00	24.00–29.75	—	28.50–31.70	28.50–32.80	28.50–31.70	28.00–32.00

Depths in meters below Land surface, EC in μS/cm at observed temperature. LN Long normal resistivity survey. SN short normal resistivity survey.

Table 2.   Depth of identified fractures zones in shallow bore wells by geophysical and geological information in Wailpally watershed, AP, India.

Bore well number	Lithologic information	EC log	Resistivity log		
			LN	SN	PR
1	Fractured granite at 50.50–59.10	57.00–60.00	55.00–58.00 61.00–62.00	54.30–55.50 58.10–59.30	56.10–57.20 59.30–61.50
2	Minor fracture at 50.20–52.40	39.00–41.00 49.00–51.00	38.30–39.70	38.10–39.20	37.30–38.70 48.70–49.10
3	Fractured granite at 45.60–48.75	44.00–46.00	46.10–47.30 48.20–49.70	45.20–46.30	47.50–48.80
4	Semi-weathered granite at 35.75–36.10 50.00–55.20	34.00–38.00 46.00–48.00 50.00–52.00	39.70–41.10	40.20–42.30	42.50–44.00

Table 3.   Depth of identified fractures zones in shallow bore wells by geophysical and geological information in Sadras watershed, AP, India.

Bore well number	Lithologic information	EC log	Resistivity log		
			LN	SN	PR
1	Fractured Charnokite at 20.75–24.10	23.00–25.00 29.00–31.00	22.30–23.80 32.50–33.60	25.70–27.00	20.80–23.70 27.10–28.30
2	Minor fracture at 24.10–28.70 40.20–44.75	25.00–27.00 29.00–31.00 41.00–43.00	28.50–30.20 40.50–42.70	30.70–32.50 43.60–45.80	21.10–24.20 43.73–46.90
3	Fractured Charnokite at 38.50–440.20	22.00–24.00 37.00–39.00	20.70–22.90 40.10–41.20	23.10–24.20 38.20–40.10	38.50–46.10 45.20–48.80
4	Semi-weathered Charnokite at 25.10–28.80 44.50–48.10	27.00–31.00 45.00–47.00	30.50–32.20 47.10–49.20	30.10–31.20 45.20–47.10	30.20–35.80

The applicability of EC logs technique has been tested, in particular in hard rock areas of India, first time. The method and their results have been validated against the results obtained by other geophysical and geohydrological techniques in these areas and in particular in above-mentioned bore wells. The geophysical techniques such as self-potential logging, point resistance logging and gamma logs[15] and hydrological techniques such as water flow measurement (injection test),[16] lithological observation, etc. were also carried out in these bore wells. Results of these studies have been compared with the results obtained by EC logs. These results very clearly show that the depth of fractures/water bearing rocks obtained by EC logs technique are more or less comparable with the above-mentioned techniques studied (Tables 1–3).

5. Conclusions

Results obtained by EC logs have been found in good agreement with other geophysical and geological findings for identifying the water bearing rock formation and fractured rock in shallow bore well of hard rock areas. This method is quick and cost effective with fairly good precision and has verified by comparing several kinds of investigations. EC logs of a bore well also supplied information about the water table (precise), temperature and electrical conductivity at different depths, information about recharge and discharge of groundwater zones, electrical conductivity at any depth and preliminary information about the chemical quality of water. In addition to this also gives useful information about the saturation, super saturation, and under saturation conditions of aquifer formation/bottom depth of the bore well.

Acknowledgments

Authors wish to thank Dr. V. P. Dimri, Director, NGRI, Hyderabad, India for his moral support and his permission to publish this paper.

References

1. D. R. Lee, *J. Hydrol.* **79** (1985) 187–193.
2. D. R. Lee, G. M. Milton, R. J. Cornett and S. J. Welch, *Proc. Second Annual Int. Conf. on High Level Radioactive Waste Management*, Vol 1, Am. Nucl. Soc., LaGrange Park, Ill, 1991, pp. 1276–1283.
3. V. K. Saxena, V. S. Singh, N. C. Mondal and S. C. Jain, *Environ. Geol.* **44**, 5 (2003) 516–521.
4. V. K. Saxena, N. C. Mondal and V. S. Singh, *Curr. Sci.* **86**, 4 (2004) 586–590.
5. P. N. Ballukraya, R. Sakthivadivel and R. Barathan, *Nordic Hydrology* **14** (1983) 33–40.
6. D. Muralidharan, *Journal of the Geological Society of India* **47** (1996) 237–242.

7. K. Subrahmanyam, S. Ahmed and R. L. Dhar, Technical Report No. NGRI-2000-GW-292, 2000.
8. R. I. Acworth, *Quarterly Journal of Engineering Geology* **20** (1987) 265–272.
9. C. F. Tsang, P. Hufschmied and F. V. Hale, *Water Resour. Res.* **26**, 4 (1990) 561–578.
10. F. E. Harvey, D. R. Lee, D. L. Rudolph and S. K. Frape, *Water Resour. Res.* **33**, 11 (1997) 2609–2615.
11. A. Revil and P. Leroy, *Geophysical Research Letters* **28**, 8 (2001) 1643–1646.
12. C. C. Satpathy, P. K. Mathur and K. V. K. Nair, *J. Marine Biol. Assoc.* **29**, 1–2 (1987) 344–350.
13. P. Sasidhar, Unpublished Ph.D. Thesis, Anna University, India, 1993, pp. 115–116.
14. V. K. Saxena and S. Ahmed, *Eviron. Geol.* **40** (2001) 1084–1087.
15. N. S. Krishnamurthy, B. Syama Prasad, G. Zamam, G. R. Anjamneyulu and S. C. Jain, NGRI Technical Report No. NGRI-2001-GW-322, 2001, pp. 1–40.
16. J. B. Charlier, NGRI Technical Report No. NGRI-September-2002, 2002, p. 54.

EXPLANATIONS OF SPOUTING DYNAMICS OF A GEYSER (PERIODIC BUBBLING SPRING) AND ESTIMATION OF PARAMETERS UNDER IT BASED ON A COMBINED MODEL COMBINING THE MATHEMATICAL MODEL (A STATIC MODEL) AND THE IMPROVED DYNAMICAL MODEL OF ONE

HIROYUKI KAGAMI*

Department of Preschool Education, Nagoya College
48 Takeji, Sakae-cho, Toyoake, Aichi 470-1193, Japan
*kagami@nagoyacollege.ac.jp

We have proposed a mathematical model (a static model), a dynamical model and a modified dynamical model of a geyser (a periodic bubbling spring) based on observation of Hirogawara geyser (Yamagata, Japan) and model experiments of the geyser. In this paper, we introduce a combined model combining the mathematical model and the modified dynamical model and report estimation of various parameters under Kibedani geyser based on comparison between results of simulation of this combined model and those of observation of Kibedani geyser. The estimation of parameters through the combined model make more reliable one.

1. Introduction

We have proposed a mathematical model,[1] a dynamical model[2] and a modified dynamical model of a geyser (a periodic bubbling spring)[3,4] based on observation of Hirogawara geyser (Yamagata, Japan)[5] and model experiments of the geyser.[1] Numerical simulations of the modified dynamical model reappear dynamics of spouting of geysers (periodic bubbling springs) and it becomes possible that parameters (surface tension on the lower interface between water and gas, volume of the underground space, depth of spouting hole and so on) under a geyser are estimated due to comparison between results of simulation and those of observation. Then we have reported trials that we compared results of numerical simulation with those of observation of Hirogawara geyser or Kibedani geyser (Shimane, Japan) and estimated parameters under Hirogawara geyser or Kibedani geyser due to the comparison.[6,7]

In the case of Kibedani geyser, a spouting mode and a pause mode are alternately observed. Drawing a graph of time variation of a position of the interface between the lump of water in the hole and atmosphere about the former mode, we can see that time taken for the position of the interface to reach a peak from the ground is shorter than one to turn back to the ground from the peak.

For reproducing these characteristic time variation of a position of the interface by the dynamical model, we additionally improved the modified dynamical model. That is, we considered that during spouting a part of gas under the lump of water was left out to atmosphere and added effects of release of gas there in proportion to length of the lump of water in a vertical direction to the former modified dynamical model. This additionally improved dynamical model enabled reproduction of characteristic time variation of a position of the interface similar to one of a spouting mode of Kibedani geyser and estimation of parameters under it.

On the other hand, a pause mode among two modes of Kibedani geyser cannot be reproduced by the dynamical model. However, it can be explained by a mathematical model (a static model).[8] Therefore, a chain of dynamics of Kibedani geyser's spouting can be expressed completely by a combined model combining the additionally improved modified dynamical model with the mathematical model. In this combined model the mathematical model or the dynamical model can have independent parameters, respectively. From this characteristics, if a value of a parameter of one model, which is decided as results of numerical simulation agrees with those of observation, is different from one of the other model, the value is regarded as unrealistic one. Therefore, concerning common parameters to both models equal value has to be had in each model. In this sense, this combined model can be considered to connote a role as verification of estimated values of parameters.

In this presentation, we will introduce this combined model and report estimation of parameters under Kibedani geyser based on comparison between results of simulation of this combined model and those of observation of Kibedani geyser.

2. Characteristics of Spouting Dynamics of Periodic Bubbling Spring

As stated above, a position of the interface between the lump of water in the hole and atmosphere of periodic bubbling spring generally varies with

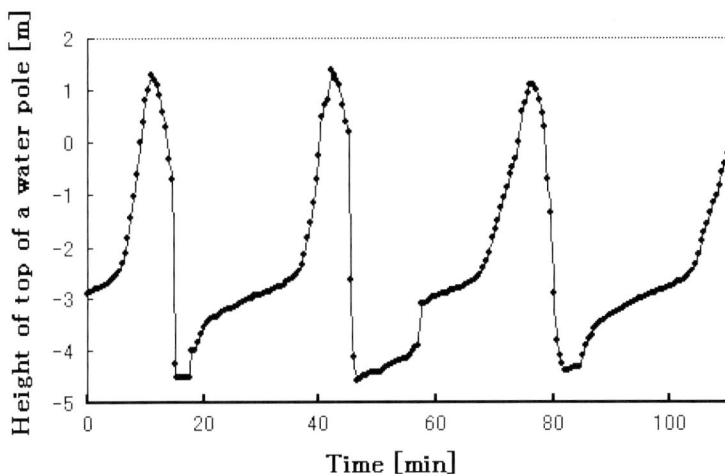

Fig. 1. Temporal variation of height of top of a water pole of Kibedani geyser (observarion) (offered by Maeda lab. at Kanto–Gakuin Univ. (http://home.kanto~gakuin. ac.jp/~kg044001/yano/sub5.shtml)).

time. At one time, the position is located above the ground. We call the period a spouting mode. Also at one time, the position is located under the ground. We call the period a pause mode. These two modes appear alternately or irregularly in obedience to periodic bubbling springs.

An example of time variation of height of top of a water pole of a periodic bubbling spring is shown below. Figure 1 shows temporal variations of height of top of a water pole of Kibedani geyser, which were observed by Maeda laboratory of Kanto-Gakuin university.[9] We can see that height of top of the water pole goes up and down by turns and a spouting mode and a pause mode appear alternately and regularly.

3. A Mathematical Model (a Static Model) of a Periodic Bubbling Spring

A mathematical model (a static model) is derived[8] based on model experiments of a periodic bubbling spring.[1] An ideal figure of model experiments of a periodic bubbling spring is shown in Fig. 2. When gas is injected sufficiently into a flask, a position of a interface between gas and water in the flask goes down under one of the lower entrance of a left spouting pipe and then water packed in the left pipe spouts owing to pressure of gas in the flask. However, water packed in the left pipe does not spout as soon as

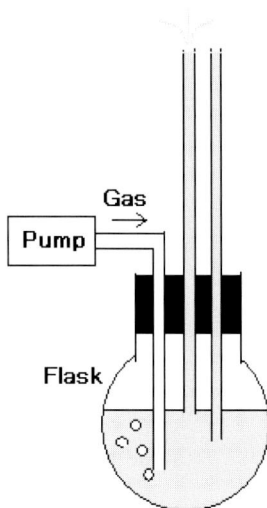

Fig. 2.   An ideal figure of model experiments of a periodic bubbling spring.

the position of the interface between gas and water in the flask goes down under one of the lower entrance of the left pipe. There is a time lag between the above two events. Concretely, a surface tension on an interface between water and gas in the lower entrance of the left pipe supports against pressure of gas in the flask for a while. So, we can write an equation of balance of pressure as

$$P_0 + \rho g H + f_k = P, \qquad (1)$$

where P_0 represents pressure of the atmosphere, ρ represents density of water, g represents gravity acceleration, H represents length of a lump of water packed in the left pipe from the lower interface between water and gas to the upper one, f_k represents pressure due to a surface tension on an interface between water and gas in the lower entrance of the left pipe, and P represents pressure of gas in the flask.

And an equation of state of gas in the flask is written as

$$P(V_0 + v) = n\alpha, \qquad (2)$$

where V_0 represents volume of gas packed over the position of the lower entrance of the left spouting pipe in the flask, v represents volume of gas packed under the position of the lower entrance of the left spouting pipe in the flask, n represents molar number of it and α is constant. Also an equation of a rate of gas supply in the flask is written as,

$$\frac{\mathrm{d}n}{\mathrm{d}t} = \beta, \tag{3}$$

where β is a constant.

Using Eqs. (1)–(3) and so on, we can write a spouting period T as

$$T = \frac{V_0}{\alpha\beta} + \frac{f_k S}{\rho g \alpha \beta}(f_k + P_0 + \rho g H), \tag{4}$$

where S represents an area of a cross section of the pipe.

4. A Dynamical Model of a Periodic Bubbling Spring

A derivation of a dynamical model and a modified dynamical model of a geyser (a periodic bubbling spring) is stated in Ref. 4 in detail. Finally, an evolution equation of temporal variations of height of top of a water pole packed in the spouting pipe of a periodic bubbling spring is written as

$$(n_0 + \beta t)(V_0 + Sx)\rho H \frac{\mathrm{d}^3 x}{\mathrm{d}t^3} + \frac{8\pi\eta H}{S}(n_0 + \beta t)(V_0 + Sx)\frac{\mathrm{d}^2 x}{\mathrm{d}t^2}$$

$$+ (n_0 + \beta t)PS\frac{\mathrm{d}x}{\mathrm{d}t} = (V_0 + Sx)P\beta, \tag{5}$$

where n_0 represents molar number of gas in a underground space just before the water pole's beginning to move up to the upper entrance of the spouting pipe and η represents viscosity coefficient. And x is regarded as a position of the lower interface between water and gas of the water pole and an upper direction of a vertical line is regarded as a plus direction of the x-axis.

5. A Combined Model Combining the Mathematical Model (a Static Model) and the Improved Dynamical Model of Periodic Bubbling Spring

A combined model consists of the mathematical model (the static model) and the improved dynamical model of a periodic bubbling spring. Concretely, temporal variations of height of top of the water pole of a periodic bubbling spring as shown Fig. 1 is dealt with by the improved dynamical model. And a spouting period, that is, time from a beginning of a pause mode to next one is dealt with by the static model.

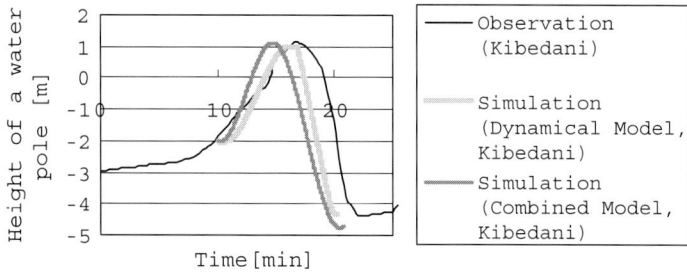

Fig. 3. Temporal variation of top of a water pole of Kibedani geyser (observation, simulations dynamical model, Combined model).

6. Comparison between Results of Simulation of the Combined Model and those of Observation of Kibedani Geyser

The combined model is first applied to Kibedani geyser because in the case of it pause modes and spouting modes appear alternately and almost regularly. Results of observation of Kibedani geyser observed by Maeda et al.[9] were used as comparison with the combined model.

From results of observation of Kibedani geyser we can see a spouting period is almost 30 min. So at first each parameter has to be decided as a spouting period $T \approx 30$ min using Eq. (4). Moreover, each parameter also has to be decided as temporal variations of height of top of a water pole of Kibedani geyser are reproduced by numerical simulations using Eq. (5).

A graph of temporal variations of height of top of a water pole obtained based on above procedure is shown in Fig. 3. And parameters used by this simulation of the combined model are as below;

$S = 0.01\,\mathrm{m}^2$, T_e (temperature of gas in the underground space) $= 320\,\mathrm{K}$, $H = 100\,\mathrm{m}$, $f_k = 22\,\mathrm{N/m}^2$, $V_0 = 990\,\mathrm{m}^3$, $\beta = 0.00019\,\mathrm{mol/s}$.

Above result of simulation is a sample. Possibly another set of parameters may bring about more suitable results of simulation. An essential point is that we can estimate more reliable parameters using the combined model.

7. Conclusions

We introduced a combined model combining the mathematical model and the modified dynamical model of a geyser (a periodic bubbling spring) and reported estimation of parameters under Kibedani geyser based on comparison between results of simulation of this combined model and those of observation of Kibedani geyser.

In the case of using only the modified dynamical model, we can select many groups of parameters suitable for results of observation. But, in the case of using the combined model, the number of groups of parameters suitable for results of observation is strictly limited because of demands from two independent models, that is, the mathematical model and the modified dynamical model. In this sense, the estimation of parameters through the combined model make more reliable one.

Acknowledgments

Results of observation of Kibedani geyser were offered by Maeda laboratory at Kanto-Gakuin university. The author would like to thank Prof. Naoki Maeda, Mr. Eiichi Ishii, and members of Maeda laboratory.

References

1. M. Katase *et al.*, Abstracts of a meeting for presenting research papers of Kanto Gakuin University College of Engineering, 1999, pp. 99–100.
2. H. Kagami, Abstracts of the 55th Annual Meeting of the Balneological Society of Japan 33 (2002).
3. H. Kagami, The 2003 IUGG General Assembly, HW04/09P/C31-004 (2003).
4. H. Kagami, *Proceedings of the 38th Conference of Sciete Internationale des Techniques Hydrothermales and the 56th Annual Meeting of the Balneological Society of Japan* (2003), pp. 55–60.
5. E. Ishii *et al.*, Abstracts of the 52th Annual Meeting of the Balneological Society of Japan 28 (1999).
6. H. Kagami, Abstract of Papers Submitted to the 56th Annual Meeting, Journal of the Balneological Society of Japan **53**, 95 (2003).
7. H. Kagami, Abstracts 2004 Japan Earth and Planetary Science Joint Meeting, H020-001 (2004).
8. H. Kagami, Abstracts of the 53th Annual Meeting of the Balneological Society of Japan **27** (2000).
9. http://home.kanto-gakuin.ac.jp/~kg044001/yano/sub5.shtml

IMPROVEMENT OF GROUNDWATER REGIME THROUGH INNOVATIVE RAINWATER HARVESTING ALONG ROAD SIDES

S. K. JAIN

Jal Bhawan, 5-Jha-2, Jawahar Nagar, Jaipur-302004
Rajasthan, India
skjain@groundwaterindia.com

The paper deals about viable and immediate solution of shortage of drinking water in the countries like India, Asian, and African continents. The paper highlights rainwater harvesting along both the sides of roads with the help of suitable, simple structures, which are easy to maintain. This may turn out to be long-term solution for the areas, which are draught prone, or having below normal rainfall. The example given in the paper for "Golden Quadrilateral" project of express national highways in India is quite illustrative and is applicable to other countries also falling in almost similar kind of climatic zones. The concept given in the paper would enhance water availability 8–10 times compared to natural process of rainfall infiltration. It would also improve quality of ground water and would save considerable energy in lifting the water due to the rise in water levels.

1. Introduction

During rainy season, most of the rain water goes as runoff and a very small portion of total rainfall (average 10%–15%) meets the underground water.[2] However, we have few perennial rivers in the north India, but with no equitable water availability in all the watersheds.[1] The situation is still worse in the western and southern India as most of the rivers are seasonal in nature. The possible solution of interlinking of rivers, desalination of seawater have to go long way to have practical benefits due to requirements of huge financial resources and likely interstate and interregional disputes.[6] Therefore, a wise, viable, and immediate solution to enhance water availability is to collect the precious water drops falling from the sky and divert most of it to ground water regime through suitable rain water harvesting system at wide scale,[7] in most parts of the country. This paper presents simple innovative methods of rain water harvesting at regional scale all over the country by suitable harvesting structures along the road sides, particularly, most of the national highways including Prime ministers Golden Quadrilateral Project Connecting whole of India north–south and west–east.

2. Methodology

The location and design of sustainable rainfall harvesting system requires hydrogeological study of the area as well as subsurface information of most permeable zone.[3] Besides, average rainfall and rainfall intensity need to be analyzed as per climatic zones. Based on normal rainfall and above normal rain fall intensity, the rain water harvesting system should be designed in such a way that 70%–80% run off is send back to groundwater regime after natural filtration process. As present paper has a limited scope of harvesting rainwater falling on the major roads, the designs are restricted to this purpose only.

3. Concept

In designing rainwater-harvesting system, capturing rainfall run off from the roads, and creating artificial connectivity to subsurface water in the hygienic manner is the key-concept. The effectiveness of the concept lies in reasonable cost, coverage of large areas, and immediate implementation and immense benefits in terms of additional water availability, improvement in water quality, increased plantation, maintaining eco-balance, etc.

4. Design

The sustainable design of rainwater harvesting along roadsides fall under two categories:

(a) The area with soil/weathered rock having vertical permeability up to water level zone. In this kind of situation, the percolation pit method would be suitable. In this method, the pits of average dimension, 3 m depth, 3 m long, 2 m wide can be made along the roads between side lanes and main road. These pits may be made at every 500-meter interval along both the sides of the road as shown in Fig. 1. The pits should have natural filtration media of coarse sand, gravels and pebbles and should be covered with perforated slabs. The road should have 1° slopes towards these pits from the divider.

(b) The area having impermeable zones prior to water level, like clays, solid rocks etc. In this type of areas, the rainwater-harvesting system

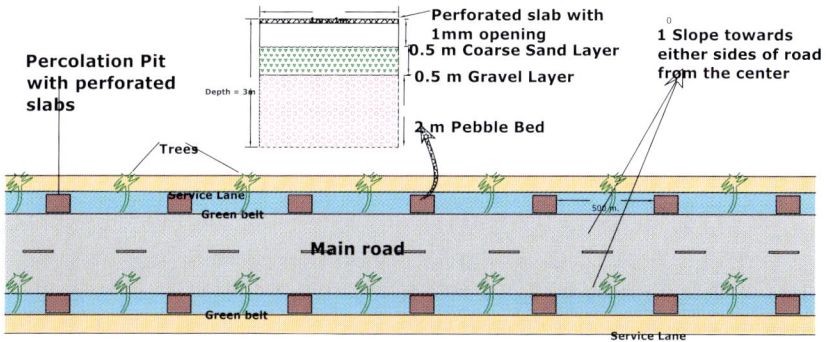

Fig. 1.   Rain water harvesting design along the roads through percolation pits.

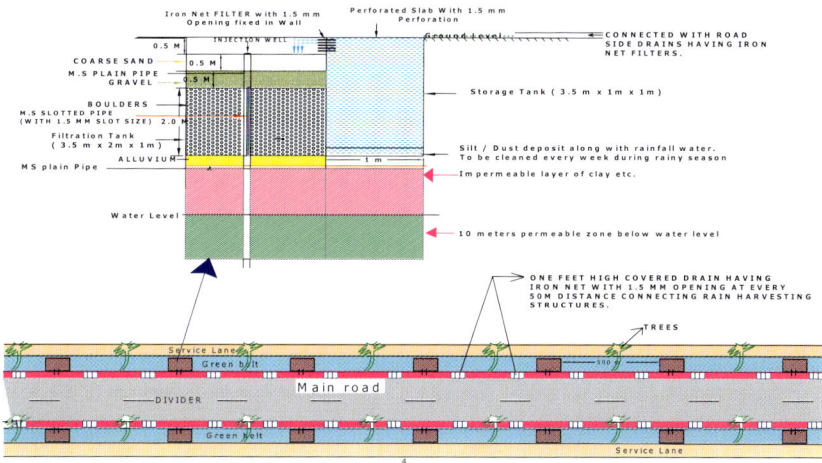

Fig. 2.   Rain water harvesting through injection well technique.

will have injection well via storage tanks and filtration tanks reaching at least 10–15 m below water level. The design is self-explanatory as per Fig. 2. Here, water is diverted to groundwater reservoir through recharge shaft via filtration media crossing the impermeable zone.

5. Advantages

(1) These systems of rainwater harvesting can be applied, in any country at massive scale, at all national highways and other roads.[4,5]

(2) The systems would add 70%–80% runoff water from roads directly to groundwater as against 10%–15% by natural process.
(3) The green belt should be developed near the percolation pits/injection wells as the trees would require less watering due to high-moisture contents near these structures.
(4) These rainwater-harvesting systems would also help in increasing the life of the roads, as there will be negligible stagnant rainwater on the roads.
(5) These would also help in creating hygienic condition, as there will be control on spread of diseases like malaria, etc. due to absence of stagnant water on the roads.
(6) The concept envisaged in this paper would be helpful in solving drinking water problems and would act as controlling factor for falling water levels, which would in turn be helpful in increasing agricultural productivity also.
(7) It would also help in saving electricity/diesel as one meter rise in water level saves about 0.4 W of electricity.

6. Limitations and Solutions

(1) The vertical iron filters to be provided in drains may be chocked due to silt/ occasional waste found on the road. The filters need to be cleaned periodically.
(2) The oil and grease from vehicles may reach in drains with water at places with heavy traffic density. The upper coarse sand layer should be replaced in filtration tanks every week during rainy season which will absorb oil and grease.
(3) Some human and animal waste materials may carry organic contamination to drains with running water. A small layer of potassium per magnate should be spread over coarse layer in filtration tanks for disinfecting water before it reaches aquifer. It should be replaced before every rainy season.

7. Example

Taking 14,000 km long "Prime Minister's Golden quadrilateral" National Highway project across India having minimum width of 30 m will have an area of 420×10^6 square meters. Presuming 500 mm as an average annual rainfall, the total runoff on roads would be $420 \times 10^6 \times 0.5 = 210 \times 10^6$ cubic meters i.e. 210 million cubic meters. By natural process, 10% of this runoff will meet the groundwater which works out to be only 21 mcm/annum; whereas, the technique given in this paper would assimilate 80% of the total runoff to groundwater which is estimated as 168 mcm/annum. This example clearly shows that 8–10 times water will be additionally stored in under ground reservoir.

8. Conclusion

The rainwater-harvesting technique presented in this paper is less expensive than any other techniques of artificial recharge, if implemented at the time of road construction itself. The innovatively designed structures are simple, easy to construct, operate, and maintain. These techniques may be initially taken up as pilot project in some priority sectors. It may turn out to be lifeline in surviving domestic water requirement during summer/draught in a matter of year or two. Besides, it may sustain water availability till the time long-term projects like interlinking of rivers desalination of sea water, etc. sees the light of the day.

Acknowledgments

The help rendered in bringing out this paper by **Anish Chatterjee**, Senior consultant, **Amit Sharma**, Manager (Technical) and **Mrs. Hemlata**, Scientific officer of GWMICC(P) LTD. Jaipur, is gratefully acknowledged.

References

1. CGWB, *Ground Water Resources of India*, Government of India Publication, 1985.
2. F. G. Driscoll, *Ground water and Wells* (Johnson Division Publication, Minnesota, USA, 1987).
3. S. K. Jain, Rooftop rainwater harvesting in Rotary Club Jaipur, GWMICC unpublished report, 2003.
4. S. K. Jain, Road side rainwater harvesting system in Rukmini Birla School, Jaipur, unpublished report, GWMICC (P)Ltd., 2004.
5. S. K. Jain, Rainwater harveting, Indo–Swis Club, Jaipur Unpublished report, GWMICC(P)Ltd., 2004.
6. K. R. Karanth, *Groundwater Assessment, Development and Management* (Tata McGrawHill Publication, India, 1987).
7. S. Ramakrishnan, *Ground Water*, (K.G. Graph Arts Publishers, Chennai, 1998).

CHARACTERIZATION OF GROUNDWATER POTENTIAL ZONES DEDUCED THROUGH THE APPLICATION OF GIS IN SEMI-ARID GRANITIC TERRAIN

V. S. SINGH* and A. K. MAURYA

National Geophysical Research Institute, Uppal Road
Hyderabad 500 007 (A.P.) India
*vssingh77@hotmail.com

In the recent years, there has been growing demand for groundwater with increase in population and developmental activities in the industry as well as agricultural sectors. In order to meet the rapid growing demand delineation of potential groundwater zone becomes essential and one of the cost effective and rapid technique is to integrate various data on geo-information through the application of GIS, followed by ground truth verification. This technique is widely being used, particularly in hard rock terrain.

In a hard rock granitic terrain, application of GIS is being made to integrate various data such as geology, drainage density, slope, lineaments and geomorphology to obtain different groundwater potential zones. In the present paper, an attempt has been made to characterize such groundwater potential zone derived through the application of GIS. The area is typically characterized by scarcity of groundwater and also affected due to high salinity and fluoride content in groundwater. In order to characterize them, various hydrogeological data such as well yield, water level and water quality collected from these zones have been analyzed.

1. Introduction

There are several methods such as geological, hydrogeological, geophysical, and remote sensing techniques, which are employed to delineate groundwater potential zones. A systematic integration of these data with follow up of hydrogeological investigation provides rapid and cost-effective delineation of groundwater potential zones. The various data are prepared in the form of thematic map using Geographical Information System (GIS) software tool. These thematic maps are then integrated using "Spatial Analyst" tool. Arithmetic, Boolean, and Relational mathematical operators are used to develop model depending on objective of problem at hand, such as delineation of groundwater potential zones. Many researchers[1-9] have done integrated study to estimate groundwater potential zones using GIS with numerous conceptual models. Khairul et al.[7] have concluded that the integration of remote sensing and GIS for groundwater potential zones is more effective in hard rock terrain but less effective in alluvium environment. Krishnamurthy *et al.*[9] have demonstrated the capabilities of remote sensing and GIS for demarcation of different groundwater potential zones, especially in diverse geological set up.

Various thematic maps such as geological, geomorphological, lineament, drainage density, and slope maps have been considered to delineate groundwater potential zones in a semi-arid zone hard rock granitic aquifer using GIS. Each of these thematic maps has been assigned suitable weightage factor. These thematic maps were then integrated using GIS based model to delineate groundwater potential zones.

In order to characterize these various zones detailed hydrogeological data were collected during the field investigation viz. well yield, water level and groundwater quality. These data were analyzed with respect to each zone to characterize them.

1.1. *Study area*

The study area (200 sq. km), a typical watershed in hard rock terrain (Fig. 1), lies in Nalgonda District of Andhra Pradesh State (India). Physiography of the area exhibits three distinct features (i) hilly terrains covering mainly northern and north-western part of the basin, (ii) mid-slope region with moderate undulating terrain, and (iii) relatively flat region with gentle slope covering larger part of the basin. The area receives an average annual rainfall (1993–2003) of about 806 mm. The surface runoff goes to river Kongal. There are several tanks across these drainages, however many of these remain dry.

1.2. *Geology*

The basin is underlained by Archaean rocks, which have suffered considerable degree of tectonic disturbances. The northwestern edge and northern part are occupied by such hills of hard crystalline massive granites. The southern fringe area is occupied with Kankar (calcium concentrate) covered with clay. These two geological units are most unfavorable from groundwater point of view. Remaining part of the area is covered with granite and gneisses. These are highly weathered, jointed, fractured and faulted, which forms potential groundwater zones. There are strips of alluvium deposits seen along the stream course, which could also be potential groundwater zones.

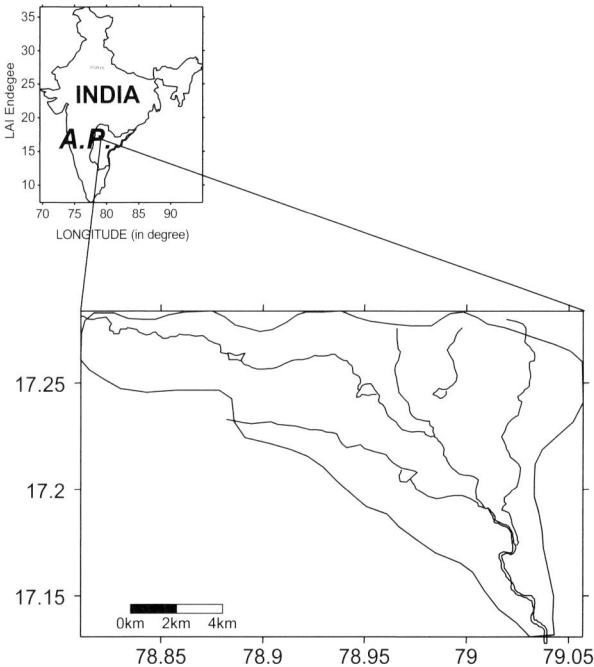

Fig. 1.   Location map of study area.

1.3. *Lineaments*

The area is cris-crossed with lineaments. The prominent directions of these are NW-SE and NE-SW. Although lineaments have been identified through out the area, it is the lineaments in the pediplain or valley fill, which are considered significant from groundwater occurrence point of view.

1.4. *Drainage*

The drainage pattern, in general, is dendritic, typical of granitic terrain. In the northern part these are dense. Many of these follow the straight course along the lineaments in the central and southern part. Some of these drainages culminate to tanks where rainwater is temporarily stored and used for irrigating crops. The drainage density in the area has been calculated after digitization of the entire drainage pattern. It varies from 0.02 to 4.85 km/sq. km.

1.5. *Slope*

Slope map of the area indicates that it varies from 0%–1% to more than 15%. We have classified the slope into five categories, i.e. 0–1, 1–3, 3–5, 5–15, and more than 15%. Most of the study area occupies slope category of 0%–1% which is favorable for groundwater recharge.

1.6. *Geomorphology*

Geomorphology reflects various landform and structural features. These units deciphered from the remote sensing data.[10] There are four major geomorphological units namely shallow weathered pediplain (PPS), shallow weathered buried pediplain (BPPS), moderately weathered pediplain (PPM) and moderately weathered buried pediplain (BPPM). The other units such as shallow weathered pediplain with alkaline soil (PPSA), rocky pediment (P), Tor complex (TC), Residual hills (RH), Denudational hills (DH), and Valley fills (VF) constitute very small part of the study area.

2. Integration through GIS

The integration of various thematic maps describing favorable groundwater zones, into a single groundwater potential map has been carried out through the application of GIS. It required mainly three steps; Spatial Database Building, Spatial Data Analysis, and Data Integration.

2.1. *Spatial database building*

Arc Catalog tools have been used to create the scheme for feature data sets, tables, geometric networks, and other items inside the database. The study area boundary, drainages, villages, rivers, streams, and ponds have been digitized from topographical map. Similarly, other maps such as geomorphological map, slope map, geological map, and lineament map have been digitized from respective maps.[10] These digitized maps have been projected in polyconic projection with respect to Everest 1956. Topology building of each map before and after projection has been carried out. Attributes to these maps have been added. Buffering of 100 m in case of lineament map has been adopted. Drainage density is calculated in "Spatial Analyst" tool.

2.2. *Spatial data analysis*

The various thematic maps as described above were converted into raster form considering 100 m as cell width to achieve considerable accuracy. These

Table 1. Weightage assigned to different features of various thematic maps.

Thematic map	Related features	Symbol	Weightage
Geology	Massive granite	P(Gr)	1
	Weathered gneiss	A(Gn)	3
	Kankar	K	1
Geomorphology	Moderately weathered buried	BPPM	3
	Shallow weathered buried pediplain	BPPS	2
	Moderately weathered pediplain	PPMA	3
	Shallow weathered pediplain	PPSA	2
	Moderately weathered pediplain	PPM	3
	Shallow weathered pediplain	PPS	2
	Denudational hills	DH	1
	Residual hills	RH	1
	Tank body	TB	2
	TOR complex with isolated hills	TC	1
	Valley fill	VF	3
	Rocky pediment	P	1
Slope	Nearly level	0%–1%	4
	Very gentle	1%–3%	3
	Gentle	3%–5%	2
	Moderate	5%–10%	1
	Moderate — steep	10%–15%	1
	Steep	15%–35%	1
	Very steep	>35%	1
Lineament	Present	—	3
	Absent	—	1
Drainage-density	Low density/coarse texture	0–1 km/km^2	4
	Medium density/medium texture	1–2 km/km^2	4
	High density/fine texture	2–3 km/km^2	1
	Very high density/superfine texture	3–5 km/km^2	1

were then reclassified assigning suitable weightage considering favorable to groundwater potentiality (Table 1).

2.3. *Data integration*

Reclassified maps were brought into the "Raster Calculator" function of Spatial Analyst tool for integration. A simple arithmetical model has been adopted to integrate various thematic maps. The final map has been categorized into five, from groundwater potential point of view, as Very Good (VG), Good (G), Moderate to Good (MG), Moderate to Poor (MP), and Poor (P).

The groundwater potential map is shown in Fig. 2. Although, the basin has been classified into different groundwater potential zones based on the

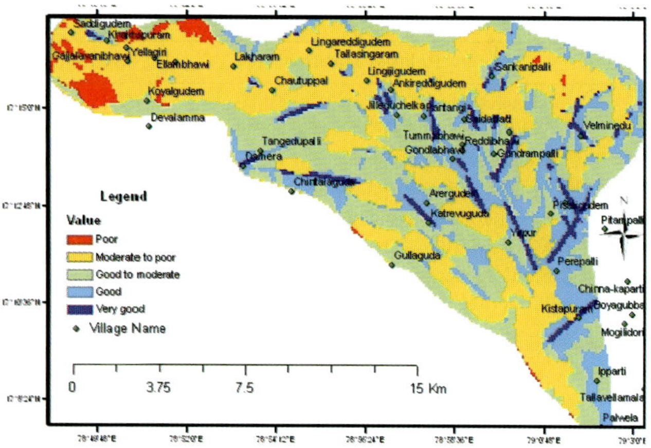

Fig. 2.   Integrated map of Kongal Basin.

cumulative effect of geology, geomorphology, drainage density, and lineaments, the final selection of drilling sites needs further probing using other
techniques such as geophysical investigations. The area classified as VG
may have very good yield. The total area is about 3% of entire basin. Similarly the area marked by G may be considered to have good yield. It is
about 18% of the total basin. The area marked by MG and MP may yield
moderately good to moderately poor. These areas occupy nearly 38% each
of entire basin. The poor prospect of groundwater is indicated by P, which
occupies about 3% of basin.

3. Characterization of Groundwater Potential Map

In order to characterize the different groundwater potential zones deciphered through GIS, hydrogeological data like well yield, depth to water
level and groundwater quality have been collected and analyzed.

Well yield: The yield of the wells have been found to vary from 699
to $10 \, \text{m}^3/\text{day}$. These values have been classified with respect to the different groundwater potential zones as described above. Although, there is no
distinguishable variation in the yield of wells; however, large percentage of
high yield wells are found in VG and G zones whereas large percentage of
low yield wells are found in MG zone.

Depth to water level: Occurrence of groundwater in various zones have
been measured in terms of depth to water level below ground level. It has

been observed that water level in the VG and G zones are shallower than the deep water level found into MP and MG.

Groundwater quality: Groundwater in the basin is typical of granitic terrain having moderately high TDS and affected due to fluoride. In order to have a broad classification of groundwater quality in the basin, electrical conductance (EC) of groundwater have been measured. These are found to vary from 760 to 4,000 μmho/cm. It has been found that fresh water of low EC occurs more in MP zone than other zones.

Similarly, low fluoride content has been found in MP zone.

4. Conclusions

The groundwater potential map derived through the integration of various information through GIS has been evaluated in terms of yield of wells, depth to water level and the quality of the groundwater. It may be concluded that:

- high yield in VG and G; low in MP;
- shallow water level in VG and G and shallow as well as deep water level in MG and MP;
- good quality of water in MG and MP zone;
- low F in MG and MP zone.

Acknowledgments

Encouragements have been made from the Director, NGRI. Valuable help has been received from Mr. M. Venkatarayudu of GIS Lab., NGRI. Authors are grateful to them.

References

1. P. K. Sikdar, S. Chakrobarty, E. Adhya and P. K. Paul, Potential zoning in and around Ranigang coal mining area, Bardhaman District, West Bengal — A GIS and Remote Sensing Approach, *J. Spatial Hydrology* **4** (2004) 1–24.
2. R. K. Jaiswal, S. Mukherjee, J. Krishnamurthy and R. Saxena, Role of remote sensing and GIS techniques for generation of groundwater prospect zones towards rural development — an approach, *Int. J. Remote Sensing* **24** (2003) 993–1008.
3. Y. S. Rao and K. D. Jurgan, Delineation of groundwater potential zones and zones of groundwater quality suitable for domestic purposes using remote sensing and GIS, *Hydrological Sciences J.* **48** (2003) 821–833.

4. A. K. Singh and S. Raviprakash, An integrated approach of remote sensing, geophysics and GIS to evaluation of groundwater potentiality of Ojhala sub-watershed, Mirjapur District, U.P., India, Map India, 2003.

5. A. K. Singh, S. Raviprakash, D. Mishra and S. Sing, Groundwater potential modeling in Chandrapraha subwatershed, U.P. using remote sensing, geoelectrical, and GIS, Map India, 2002.

6. P. Lachassagne, R. Wyns, P. Berard, T. Bruel, L. Cherry, T. Coutand, J. F. Desprats and P. L. Strat, Exploration of high yields in hard rock aquifers: downscaling methodology combining GIS and multicriteria analysis to delineate field prospecting zones, *Groundwater* **39** (2001) 568–581.

7. K. A. Musa, J. M. Akhir and A. Ibrahim, Groundwater prediction potential zone in Langat Basin using the integration of remote sensing and GIS, ACRC, 2000.

8. S. Sahid and S. K. Nath, GIS integrated of remote sensing and electrical sounding data for hydrogealogical exploration. *J. Spatial Hydrology* **2** (2001) 1–12.

9. J. Krishnamurthy, N. K. Senates, V. J. Raman and M. Manvel, An approach to demarcate groundwater potential zones through remote sensing and geographical information system, *Int. J. Remote Sensing* **17** (1996) 1867–1884.

10. Andhra Pradesh Remote Sensing Application Agency (APSRAC) — Geological, Geomorphological and Slope maps of Nalgonda District, A.P., 1992.

WATER DEPLETION OF FOUR SOIL LAYERS IN THE TROPICS

HARIS SYAHBUDDIN[*,‡,§] and MANABU D. YAMANAKA[*,†,¶]

Graduate School of Science and Technology Kobe University, Japan
†*Institute of Observational Research for Global Change (IORGC)*
Japan Agency for Marin-Earth Science and Technology (JAMSTEC), Japan
‡*Indonesia Agro-climate and Hydrology Research Institute (IAHRI), Indonesia*
§*haris@ahs.scitec.kobe-u.ac.jp*
¶*mdy@kobe-u.ac.jp*

Due to the contribution of water from land surface to the atmosphere is pre-dominant, therefore soil water depletion analysis is necessary. This study was conducted based on the principle of soil water budget. The soil water depletion in the day without rain is dominated by water losses upward and opposite in the day with extreme rain.

1. Introduction

Many scientists consider, the biggest part of water transport to the atmosphere is provided by land surface through an evaporation (E) and transpiration (Tr). To study the contribution of water from land surface, we have to conduct research about the soil water budget at the first place. However, most of study result about water budget or moisture balance which have relationship with global climate phenomenon,[1] water and heat flux,[2] water storage in basing,[3] and crop requirement and water use,[4,5] were reported only for top soil layer or interaction between soil surface and atmospheric boundary layer (ABL). On the others hand, some studies of water budget which are consider about soil layer do not include some important factor, such as characteristic in each a soil horizon and vegetation in land surface (surface roughness).[6]

The propose of this paper is to explore a soil water depletion (SWD) from four soil layers based on an actual environment of soil horizon characteristic, where saturated soil hydraulic conductivity (Ks) and soil depth is varies for each soil type and layer.

2. Model Proposed

2.1. Basic concept

Allen et al.[7] present a basic equation of SWD model firstly only for surface layer. It is not adequate enough to calculate the SWD from layer to layer. Briefly speaking, the SWD submodel in this paper was developed based on two principles of water movement in soil profile. The first is the water that goes through the soil downward after rainfall (R) or irrigation (I) applied. In this phase, the water fills top profile at first. Gravitationally, it will not move to second layer before all of soil spaces (pores) filled by the water or saturation. The other one is the water goes through the soil upward affected by E and Tr. In this phase, the water from lower layer will move and remain to upper layer capillary. In both conditions saturated and unsaturated phases profile will be find that they are not dependent on processes each other.[8]

Other importance factor is a rooting distribution or root density, which its extraction gives impact to water movement and show an exponential relationship with soil depth.[9-11] An effect of root density to Tr and water movement might be integrated by using root form type presented by Kutschera and Lichtenegger.[12] In this model, coefficient partial volume four root types were adopted (Fig. 1).

Precentage of root volume (%)

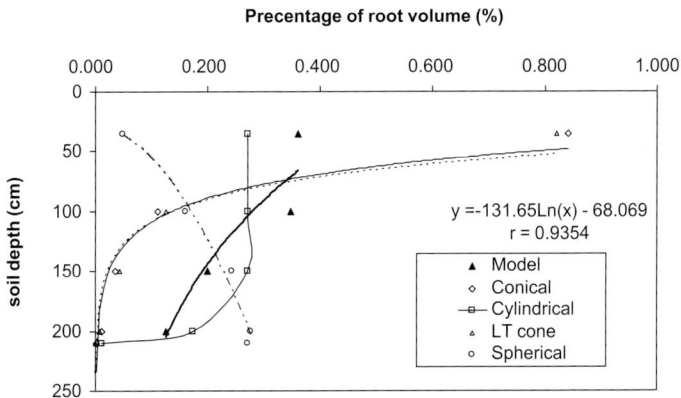

Fig. 1.   The percentage of mean value of partial volume for each root crop to total root volume, the coefficient 0.4 for the top soil layer, 0.3 for the second, 0.2 for the third, and 0.1 for the fourth layer were found. To simply an effect of root density to water movement it is expressed by multiple of Tr with those coefficients according to each layer, where its decrease exponentially for soil depth. LT cone is like a truncated cone.

To obtain the SWD computation accurately, some assumptions were implemented consciously to a characteristic of soil profile. Those are: (1) initial SWD is equal to a delta of soil water content ($\Delta\theta = \theta_t - \theta_{t-1}$), (2) surface runoff ($RO$) is zero, (3) water quantity from I and loss through drainage is zero, (4) soil and ABL layer are as the thing which are closed water column system, (5) when rainfall occurs and the water fill in soil pores dominantly, then the air is suppressed and SWD upward is zero, and (6) to conserve a water balance in the third layer then an initial SWD in the fourth layer is equal $0.75\,\Delta\theta$ of the third layer.

According to all of those principles, the equations of SWD from four soil layers are written as follows:

$$De_t(1) = De_{-(1)t-1} - \Delta De_{+(2)t} - R_t + \frac{E_t}{f_{ew}} + 0.4Tr_t + \Delta DPc_{-(1)t}, \quad (1)$$

$$De_t(2) = De_{-(2)t-1} - \Delta De_{+(3)t} + 0.3Tr_t - \Delta DPc_{+(1)t-1} + \Delta DPc_{-(2)t}, \quad (2)$$

$$De_t(3) = De_{-(3)t-1} - \Delta De_{+(4)t} + 0.2Tr_t - \Delta DPc_{+(2)t-1} + \Delta DPc_{-(3)t}, \quad (3)$$

$$De_t(4) = De_{-(4)t-1} + 0.1Tr_t - \Delta DPc_{+(3)t-1} + \Delta DPc_{-(4)t}, \quad (4)$$

where De is cumulative SWD (mm), I is irrigation infiltrates the soil (mm); R, RO and E ($E = ET_oK_e$) are in mm, Tr is in an exposed and wetted fraction of the soil surface layer (mm) ($Tr = ET_oK_{cb}$).[7] DPc is deep percolation. f_w is fraction of soil surface wetted by irrigation or rainfall and without its $(0.01-1)$, and f_{ew} is exposed and wetted soil fraction $(0.01-1)$. It is calculated as $f_{ew} = \min(1-f_c, f_w)$, where fc average is exposed soil fraction covered by vegetation $(0.01-1)$. $\Delta De_{+(n)t} = De_{+(n)t} - De_{+(n)t-1}$ and $\Delta DPc_{-(n)t} = DPc_{-(n)t} - DPc_{-(n)t-1}$. The negative value of $\Delta DPc_{-(n)t}$ is not allowed. $DPc_{-(n)t}$ its self is $R_t - De_{t-1}$. The water infiltrate will enter to second layer if $(1 - (DPc_{t-1}Ks^{-1}))$ top layer is > 1. It means that, all of soil pores in each layer had been filled by water and saturate when depth of water percolation is equal to Ks.

3. Soil Water Content Fluctuation and Water Depletion

Table 1 shows a general $\Delta\theta$ in soil layer from dry to wet pattern. In order to model requested the initial SWD, a fluctuation of θ between losses (-1) and gain water $(+1)$ in three soil profiles are proposed.

In order to elaborate an important factor which has effect toward SWD significantly such as a climate, crop type, and Andosol soil parameters from

Table 1.   Pattern of delta soil water fluctuation in three soil profiles.

Soil layer	Dry pattern				West pattern			
	H-03	H-12	H-13	H-23	S-12	S-13	S-23	S-03
First	−	−	−	+	+	+	−	+
Second	−	−	+	−	+	−	+	+
Third	−	+	−	−	−	+	+	+

Note: H-03: water losses from all of layers; H-12: from the first and second layer; H-13: from the first and third layer; and H-23: from the second and third layer. S-12: gain water into the first and second layer; S-13: into the first and the third layer; S-23: into the second and third layer; and S-03: all of layers gain water.

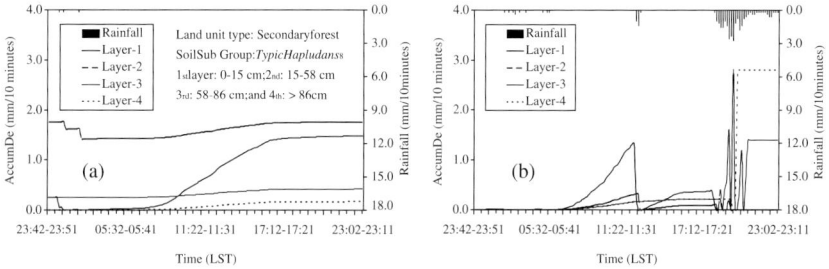

Fig. 2.   In case of H-03 at July 17, 2004, where ETc was lowest than other days and the total ΔDe was about 1.582 mm/day. (a) Nevertheless, in case of S-03 followed by extreme rainfall at November 8, 2003, the total ΔDe was around 7.915 mm/day. (b) In dry season ΔDe was dominated by upward SWD, but in the wet season, it was opposite.

Kototabang site, west Sumatra, Indonesia, was utilized.[13] The climatic wet (November 7–8, 2003) and dry season (July 17–18, 2004) were examined. In general ΔDe is from top layer for both seasons and all $\Delta\theta$ fluctuation patterns (see Table 1) always larger than others, which it was around 0.483–3.866 mm/day except for H-03 and H-13 pattern at July 17, 2004. The H-03 shows that remain water from the third and the fourth layer are not quiet enough to compensate water lose in the second layer, especially in the cloudy day and there is no rain (Fig. 2(a)).

3.1. Water depletion in case of without or light rain

Figure 2(a) also proved that in case H-03 when rainfall appeared around 0.6 mm/day in the night, water touches the topsoil in the $\Delta\theta$ negative fluctuation. It will suppress directly in certain minute water movement upward. However there is a time lag in the second layer between first time rainfall touches soil and the decreasing of SWD. Therefore, ΔDe was from the second layer (0.789 mm/day) always higher than topsoil (0.483 mm/day). In case of H-13, the rainfall water suppressed the SWD upward from top layer around mid night and the SWD from the second layer to become hampering. The water rainfall depresses not effective for the third layer. Due to the $\Delta\theta$ still negative, ΔDe from the second layer $<$ the third layer, ΔDe profiles are same with case S-12. By applying of H-23 pattern shows a consistency of soil water availability impact to reduce De upward when the rainfall appeared. Because the $\Delta\theta$ in topsoil is positive, the water rainfall fills pore soil more deeply, and then De from the second layer was affected.

3.2. *Water depletion pursue with heavy rainfall*

In case of November 8, 2003 when the heavy rainfall occurs in the night (28.4 mm) in case of all $\Delta\theta$ patterns, the water percolate and reach the fourth layer as downward SWD and dominate ΔDe into deep layer (Fig. 2(b)). By using the initial SWD are as S-13, S-23, and S-03, the amplitude of ΔDe profiles on wet season was similar. It was around 5.955 mm/day, and the level of De is not in order, where it was in the first layer $>$ the second layer, and then the second layer $<$ the third layer $<$ the fourth layer.

4. Conclusion

There are eight possibility of $\Delta\theta$ fluctuation patterns in three soil profiles which it range from dry (H-03) to wet (S-03) condition. By applying those patters as initial SWD into model found that, the first layer SWD is always exist and higher than others layer, except for H-03 followed on by cloudy day with minimum threshold rainfall is around 0.6 mm/day, and H-13 and S-23 pattern for extreme rainfall. The total of ΔDe is more elevated when the positive $\Delta\theta$ is predominant. The SWD in the day without rain is dominated by water losses upward and opposite in the day particularly with extreme rain.

Acknowledgments

The climatologically data was provided from Dr. Hiroyuki Hashiguchi from Radio Science Center for Space and Atmosphere, Kyoto University, Kyoto, Japan. This study is also supported by IORGC JAMSTEC, Japan.

References

1. A. J. Pitman, M. Zhao and C. E. Desborough, *J. Meteor. Soc. Japan* **77**, 1B (1999) 281–290.
2. H. Douville, E. Bazile, P Caile, D. Giard, J. Noilhan, L. Peirone and F. Taillefer, *J. Meteor. Soc. Japan* **77**, IB (1999) 305–316.
3. Z. Cao, M Wang, B. A. Proctor, G. S. Strong, R. E. Stewart, H Ritchie and J. E. Burford, *Atmosphere-Ocean* **40** (2002) 2.
4. P. N. J. Lane, J. Morris, Z. Ningnan, Z. Guangyi, Z. Guoyi and X. Daping, *J. Agri. Forest Meteo.* **124** (2004) 253–267.
5. A. I. J. M. van Dijk, L. A. (Sampurno) Bruijnzeel and J Schellekens, *J. Agri. Forest Meteo.* **124** (2004) 31–49.
6. J. Xu and S. Haginoya, *J. Meteor. Soc. Japan* **79**, 1B (2001) 485–504.
7. R. G. Allen, L. S. Pereira, D. Raes and M. Smith, *Crop Evapotranspiration Guidelines for Computing Crop Water Requirement* (FAO Paper56, Italy, 1998).
8. D. Hillel, *Introduction to Soil Physic*, Department of Plant and Soil Sciences, University of Massachusetts, MA, 1996.
9. E. Tasser and U. Tappeiner, *J. Ecol., Model* **185** (2005) 195–211.
10. R. Van de Moortel, J. Deckers and J. Feyen, Groei en vitaliteit van wintereik en zomereik in relatie tot de waterhuishouding, Internal Pub. No.53, Institute for Land and Water Management, Leuven, Belgium, 1998.
11. T. M. Wynn, S. Mostaghimi, J. A. Burger, A. A. Harpold, M. B. Henderson and L. A. Henry, *Variation in Root Density along Stream Banks* (Technical Report, Virginia, 2004).
12. L. Kutschera and E. Lichtenegger, *Wurzelatlas mitteleu-ropaischer Grunlandpflazen, Band 1 Monocotyledonea* (Gustav Fishcer, Stuttgart, 1992), p. 516.
13. H. Syahbuddin, M. D. Yamanaka, E. Runtunuwu, Sawiyo and T. N. Wihendar, *Impact of Soil Water Transport to Water Budget in the Atmospheric Boundary Layer at Kototabang, West Sumatra, Indonesia*, AOGS-second Annual Meeting, June 20–24 2005 (Abstract listing, Singapore, 2005).

CHARACTERISTICS OF SOIL WATER MOVEMENTS AND WATER TABLE AT THE LEIZHOU PENINSULA, GUANGDONG PROVINCE, CHINA

CHANGYUAN TANG[*,¶], JUN-HONG CHEN[†],

AKIHIKO KONDO[‡] and YINTAO LU[§]

[*]Faculty of Horticulture, Chiba University, Chiba Prefecture, Japan
[†]Guangzhou Institute of Geography, Guangzhou, China
[‡]Center for Environmental Remote Sensing, Chiba University, Chiba, Japan
[§]Graduate School of Science and Technology, Chiba University, Chiba, Japan
[¶]cytang@faculty.chiba-u.jp

Knowledge of the mechanisms of water at tropical red soil is the key to understand the hydrological cycle in south part of China. In order to study water infiltration and water table responding to rainfall events, field experiments were conducted in the Leizhou Peninsula, China, since June 2000. TDRs were set at 10, 30, 60, 100, and 200 cm in depth, respectively. Water table and rainfall were also measured at the same time. All items were recorded automatically at the intervals of 10 min. Results show that (1) there is yearly periodic variation of groundwater table at the Leizhou Peninsula, with an ascending period from about June to October, lasting for 90–100 days, and descending period during other times; (2) in the recharging area, the variation of groundwater table has close relationship with rainfall, of which during ascending period there is an obviously positively interrelationship between accumulated ascending-range of groundwater table and accumulated rainfall, and during dry reason with less rainfall; and (3) soil water infiltration responded rapidly to rainfall events with a little overland flow. Finally, based on the results water resource development in the Leizhou Peninsula has been discussed.

1. Introduction

Land surface hydrologic processes play an important role in the global water cycle. The temporal and spatial distributions of storages and flows within a catchment depend on the complex interaction of topography, geology, soil types, land use and the variability of meteorological factors.

Knowledge of the factors controlling storage and movement of water in the soil is important in order to understand a wide range of processes of hydrological importance. The initial soil moisture conditions significantly influence the runoff volumes produced during a heavy precipitation but long records of these measurements are not easy to obtain especially in rough topographic conditions. Findell and Eltahm[1] analyzed direct observations on soil moisture and rainfall and found a significant lag correlation between soil moisture and rainfall.

The soil moisture content before precipitation is crucial in estimating rainfall-runoff. In fact, the ability of soil to store water and the rate of storage depend primarily on the initial soil moisture itself and thus on the previous evaporation and drainage.[2]

This study has been prepared as a part of wider research carried out over recent years in order to assess water resources in Leizhou Peninsula, the tropic region of China. As, in this region, aquifers provide the principal water supply, one of the significant problems was to evaluate rainfall as the only source of groundwater recharge and to research the mechanism of aquifer replenishment. Therefore, the objective of this study was to examine (1) response of soil water responding to rainfall in the red soil; (2) the long term variations of water table by considering recharge processes of precipitation to groundwater; and (3) the water balance in the study area.

2. Geographical Setting and Methods

Located at the south part of Leizhou Peninsula, Guangdong Province, China, the hydrological experimental station ($110°10'00''$E, $20°19'29''$N) has mainly got a warm and humid tropical climate. This typical marine monsoon weather features a rainy season from April to September with frequent typhoons (Fig. 1). It has a rich average rainfall of 1,200–1,800 mm annually with 80%–90% falling during the rainy season. Geologically, old

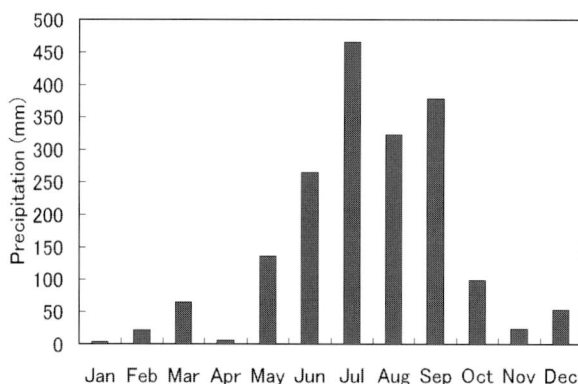

Fig. 1.   Average precipitation in the study area (1957–2002).

volcano lavas with high permeability distribute widely in Leizhou Peninsula, forming very good hydrological conditions for groundwater. In the study area, the surface was covered by 2.5–3 m thick of red soil underneath the basalt. Since the red soil has a high hydraulic conductivity ranging from 6.16×10^{-5} and 1.11×10^{-2} cm/s, and crack often can be found at the surface when the soil at top layer dries up in the shiny days, rain water can very easy to infiltrate to the ground. As a result, there is no river in the study area and groundwater is the only source for agriculture and drinking. Its annual average temperature is over $23°C$. In dry season (from October to March), water shortage problem is very serious since the high evaporation and wilted point occurred often at the top soil layer. The height of the study area is 80 m above the sea level.

Measurements for soil water, groundwater as well as overland flow have been conducted in four year period start from June 2000. In order to measure the water balance in the study area, five TDR sensors (CS610, Campbell) have been set at the depth of 10, 30, 60, 100, and 200 cm, respectively. At the same time, tensiometers were set at the same depths as the TDR sensors. Rainfall was measured by rain gauge. All data of rainfall and TDRs were recorded in a datalogger CR10 (Campbell) with the interval of 10 min. Also, the overland flow and the water table were measured in the interval of 10 minutes.

3. Results and Discussions

3.1. *Variations of soilwater during the events*

Figure 2 shows the vertical profile of soil water change during July 25–29, 2001 with total rainfall of 261 mm, where we set 0:00 on July 25 as the starting point. There was a rain from 28 h to 33 h 50 min. Before the rain event, soil volumetric water content at 10 cm in depth was 30.2%, and responded quickly to the rain. It changed to 48.3% at 28 h 50 min and reached the peak of 67.5% at 30 h 30 min. The changes of soil volumetric water content at 30, 60, 100, and 200 cm in depth began at 29 h 30 min, 30 h 20 min, 30 h 30 min and 30 h 40 min, respectively. As a result, soil water increased at all depths and had their maximum water contents ranging from 63.8% to 72.8% during the rainfall period. After the rain, soil volumetric water content at 10 cm in depth decreased continually, and reached to 50.2%, 40.1%, and 30.9% at 36 h, 64 h, and 144 h, respectively. However, soil volumetric water content at other depths returned to their values before the event at about 130 h. During this event, overland flow occurred.

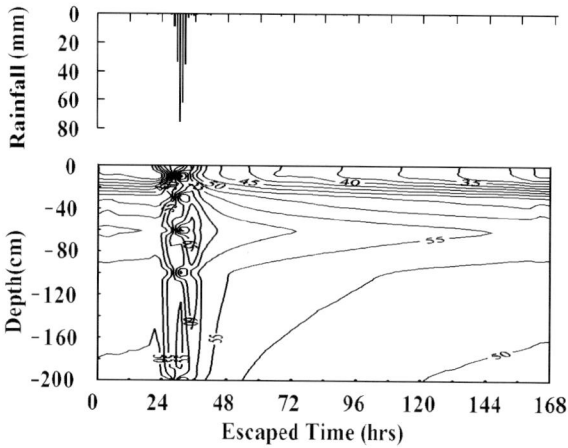

Fig. 2.   Vertical variations of soil water content from July 25 to 29, 2001.

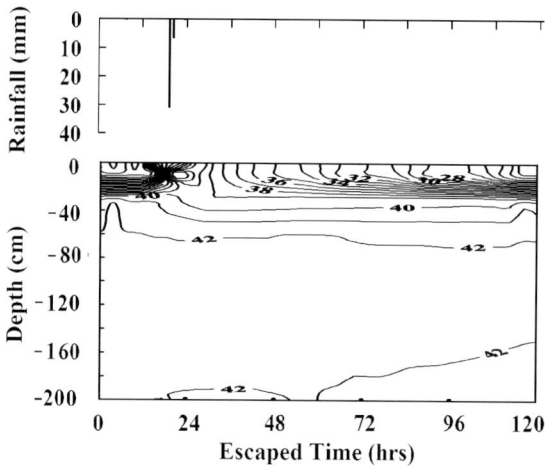

Fig. 3.   Vertical variations of soil water content from May 3 to 7, 2001.

Figure 3 shows the variations of soil water from May 3 to 7, 2001. In this case, we set 0:00 on May 3 as the starting point for convenient. During this period, there was 53 mm rainfall in total from 19 h to 20 h 30 min. The soil volumetric water contents at 10, 30, 60, 100, and 200 cm in depth were 22.8%, 40.9%, 41.9%, 42.8%, and 41.9%, respectively. Soil volumetric water content at 10 cm in depth began to increase at 19 h 40 min and became 53.4% at 19 h 50 min when the rain reached its peak. After the rain, soil volumetric water content at 10 cm in depth decreased continually, and returned to the value before the rain event about 112 h. However, soil volumetric water content at other depths changed little during the event, which indicates no rainwater recharge for the groundwater in this case. Also, there was no overland flow observed during the event.

Comparison of Figs. 2 and 3, it was found that the soil volumetric water content at 10 cm in depth was always low in the shiny day, which means that soil water at top layer loses easily because the strong potential evaporation in the study area. At the same time, soil volumetric water content at the top layer responded rainfall well during the rain event. It took about 40–50 min for rainfall infiltrated from the surface to the soil layer 10 cm in depth.

3.2. Soil water movement effected by drying/wetting at the top layers

Table 1 shows the changes soil volumetric water contents with depth during the study period. Water in the soil resides within soil pores in close association with soil particles. The largest pores transport water to fill small pores and soil particles. After a complete wetting and time is allowed for the soil to dewater the large pores, red soil in the study area has an half of the pore space for water and the rest for air. Generally, this condition is called field capacity ranging from 34% to 41% in the study area (Table 1).

Evapotranspiration, or evaporation if on a bare soil, is mainly driven during the first stage by atmospheric conditions (radiation, wind speed, etc.); during this stage the soil is wet and conductive enough to supply water at a rate commensurate with the evaporative demand. During the second stage (usually longer than the first one)[3-5] the evaporation rate falls progressively below the potential rate; at this stage the evaporation rate is limited by the rate at which the drying soil profile can deliver moisture toward the surface (i.e. by the hydraulic soil characteristics). Therefore, soil moisture is a crucial link of the air–soil interaction.

Table 1.   Changes of soil water contents with depth during the study period.

Depth (cm)	10	30	60	100	200
Minimum of soil water content (%)	14	24	34	34	34
Maximum of soil water content (%)	73	72	70	68	65
Field capacity (%)	34	35	39	41	40

The state of the soil moisture, as described by the level of saturation relative to the soil field capacity, is regulated by rainfall and potential evaporation. Both of these atmospheric driving forces exert significant control on the evolution of the soil moisture state and appear explicitly in the soil water balance equation. On the other hand, the level of soil saturation determines the availability of water as well as the hydraulic properties of the soil; thus, soil saturation exerts significant control on the rates of exfiltration and subsequent evaporation.

In the shiny days, the topsoil dried up quickly and formed the dry surface layer (DSL). Liquid water transport from deeper soil layers stops at the bottom boundary of the DSL, in turn vapor water transport is dominant in the DSL. As a result, the effect of soil moisture deficit in the top layer on the restriction of evaporation from soil surfaces of the study area kept the soil water changing little at the soil below 30 cm in depth. Yamanaka and Yonetani[6] showed a conceptual model to simulate the dynamics of the evaporation zone and those related to the form of water content profile in dry sandy soils, which can explain the difference in the surface resistance and soil water content of surface soil relationship between interdiurnal and diurnal time scales.

On the other hand, drying/wetting process creates cracks in the top soil that forms the DSL and prevents the soil water loss at the underlaid layer from evaporation in the study area. When the soil is comparatively dry, the cracks are empty and do not contribute to the water flow. At moderate rate of rainwater input to the ground surface, the cracks will rapidly be emptied by flow through the walls in response to capillary potential gradients to the surrounding soil matrix. But at high rates of rainwater inflow and/or small saturated hydraulic conductivity of the soil matrix caused by sealing during the rain event, saturated or near saturated zones built up around the cracks, which then remain water-filled and made a major contribution to the flow. When the rain continued, the cracks disappeared since the soil particles filled full of the cracks. As a result, overland flow occurred and soil matrix became the one factor to control infiltration process in the field.

3.3. *Variations of long-term variations of water table*

Figure 4 shows the variations of water table and rainfall during 2001–2002 at study area. It was found that the water table depressed continually at the early of rainy season. The water table began to rise on June 30, 2001 and August 2, 2002, respectively. In spite of rain patterns were different for these

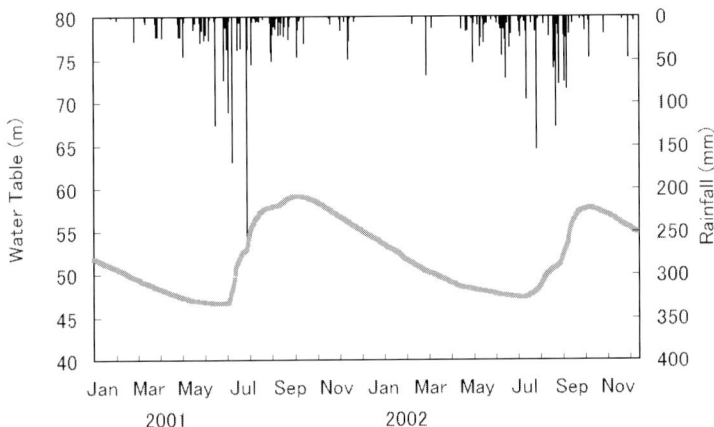

Fig. 4. Variations of water table and rainfall at study area during 2001–2002.

2 years, the accumulated rainfalls from the beginning of year were 720.8 mm in 2001 and 715.8 mm in 2002 before the water table rising, respectively. The water table kept rising and reached its peak values of 59.08 m above the sea level on September 27, 2001. However, the water table got its peak of 57.75 m above the sea level on October 29, 2002. Clearly, there was a time lag between the rainy season and the rising of the water, which was dependent on the accumulated rainfall rather than the starting date of rainy season. The water table reached its peak at the beginning of the dry season.

Figure 5 shows the relation between rainfall and changes of water table at study area during 2001–2002. It was found that the water table responded to rainfall well in the period from the date beginning to rise to the end of rainy season. As a result, this period should be considered as recharge period for groundwater by the rainfall in the study area.

3.4. *Water balance in the study area*

Precipitation falling within the study area can be stored there, return to the atmosphere as evapotranspiration or be transported out of the study area as runoff or recharge to groundwater. This can be expressed by the water balance equation as the following.

$$P = E + Q + R + \Delta S \qquad (1)$$

where P is the rainfall, E the evapotranspiration, Q the overland flow, R the recharge for groundwater, and ΔS the change in storage for soil

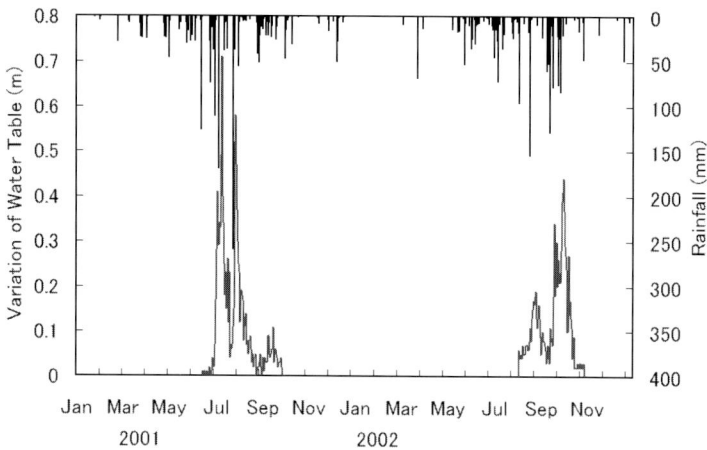

Fig. 5. Relation between rainfall and changes of water table at study area during 2001–2002.

Table 2. Comparison of water budget of 2001 and 2002 at the study area (mm).

Year	Rainfall	Evaporation	Overland flow	Recharge to groundwater
2001	2074	797	165	1111
2002	1850	665	183	1002

water (which can be either positive or negative). Over a long period, ΔS can be negligible compared to the other terms in the equation. Then Eq. (1) may be simplified to the static water balance equation:

$$P = E + Q + R. \tag{2}$$

Supposed the groundwater recharge is available in the period of the water table rising, the precipitation in the other period should be considered as evaporation which contributes nothing for the groundwater. As a result, Eq. (2) can be used to assess groundwater recharge in the study area based on the measurements of precipitation, overland flow. Table 2 shows the water budget of 2001 and 2002 at the study area. In spite of difference of annual rainfall between 2001 and 2002, 53.56% (2001) and 54.16% (2002) of annual rainfall recharged to groundwater. The soil moisture change after a precipitation event depends on both the amount of water precipitated and the value of the soil moisture before the precipitation. Excluding the heaviest cases, the experimental data presented here indicates that the soil moisture condition before the event is more relevant than the precipitation. Therefore, the monitoring of soil moisture at the top layer could permit to predict possible runoff conditions, which could also cause flooding events. At the beginning of dry season, soil water content can be high as a result of rainy season. However, the soil is commonly depleted of soil water at the end of dry season.

4. Conclusions

The following conclusions can be drawn.

(1) Evaporation, overland flow and recharges to groundwater were about 37.3%, 8.9%, and 53.8% of annual rainfall in the study area, respectively.

(2) Since the cracks and high permeability of red soil in the study area, no overland flow happened with daily rainfall less than 60 mm/day.

(3) There is a time lag between beginning of rainy season and rise of the water table. The delay for ascending period for the water table was largely dependent up on the accumulated rainfall in the year. However, the descending period of the water table started at the end of rainy season.

(4) The soil water content in the topsoil changed greatly because of strong evaporation in the shiny day. The dry layer and cracks forming in the surface help to prevent soil water loss from evaporation in the soil layer 30 cm below the surface. Also, cracks and surface sealing at the top soil during the rainfall event should be the key for understanding overland flow occurring in the study area.

References

1. K. Findell and E. A. B. Eltahm, An analysis of the soil moisture-rainfall feedback, based on direct observations from Illinois, *Water Resour. Res.* **33**, 4 (1997) 725–735.
2. K. M. Loague and R. A. Freeze, A comparison of rainfall-runoff modeling techniques on small upland watersheds, *Water Resour. Res.* **21** (1985) 229–248.
3. D. Hillel, Simulation of evaporation from bare soil under steady and diurnally fluctuating evaporativity *Soil Sciences* **120**, 3 (1975) 230–237.
4. D. Hillel, On the role of soil moisture hysteresis in the suppression of evaporation from bare soil under diurnally cyclic evaporativity, *Soil Sciences* **122**, 6 (1976) 309–314.
5. M. Menziani, S. Pugnaghi, L. Pilan, R. Santangelo and S. Vmcenzi, Field experiments to study evaporation from a saturated bare soil, *Physics and Chemistry of the Earth (B)* **24**, 7 (1999) 813–818.
6. T. Yamanaka and T. Yonetani, Dynamics of the evaporation zone in dry sandy soils, *Journal of Hydrology* **217**, 1–2 (1999) 135–148.

GROUNDWATER RECHARGE EVALUATION IN A RIVER BASIN, ANDHRA PRADESH, INDIA

RAMESH CHAND*, N. C. MONDAL, V. S. SINGH and A. PRASANTI

National Geophysical Research Institute, Uppal Road, Hyderabad-500 007, India
*rameshtyagi@yahoo.com

The average natural recharge to the phreatic aquifer of a river basin, Andhra Pradesh, India, is estimated using injected tritium technique. Tritiated water was injected at 46 selected sites in this basin before the onset of monsoon covering different soil types. Moisture content and tritium activity of the soil core samples, collected from the injected sites after the monsoon rains were measured. The variation in tritium activity and moisture content with depth is used for the estimation of tracer movement and calculation of recharge. The recharge values were found varying from 26 to 94 mm with the mean value of 29 mm corresponding to the rainfall of 615 mm. The quantum of groundwater recharge through vertical infiltration has been estimated as 211 million cubic meters over an area of $8{,}650\,\text{km}^2$.

1. Introduction

Kunderu River basin located in Anantapur, Cuddapah, and Kurnool districts of Rayalaseema region of Andhra Pradesh (India) was selected for the estimation of recharge potential. The climate is semi-arid and this basin is chronically drought affected due to frequent failure of monsoons in the last decade. The surface water resources are limited to the quantum supplied through the Kurnool–Cuddapah canal in the Kunderu basin. There were curtailments in these supplies too, owing to depletion in reservoir storage. As such large areas of the basin depends only on the groundwater supply. The prime source of recharge to groundwater in the basin is by direct precipitation due to annual rainfall. Due to non uniform recharge conditions and precipitation, large fluctuation in groundwater levels are often observed at many places. It has, therefore, become essential that the annual input or recharge to various aquifers of this basin is quantified and the data is used for planning and optimal utilization of the groundwater resources.

2. About the Study Area

Major part of Kunderu basin is located in Kurnool district of A.P. and a minor part of the lower reaches fall in Cuddapah district of A.P. The basin is bounded within latitudes $14°30''–16°$N and longitudes $77°45''–79°$E. The total area of the basin is about $8,650\,km^2$ comprising a number of sub basins drained by the Vakileru, Palleru, Galeru, Kollamvagu, Jeruru, Palieru, Tigaleru, and Maddieru streams. The key map showing the location of Kunderu basin is shown in Fig. 1. Kunderu river is the tributary of river Pennar. Figure 1 shows the major tributaries of Pennar river and the location of the basin. Basin shows both dendritic and parallel to sub-parallel drainage pattern. The basin receives both SW and NE monsoons but do not enjoy the full benefit of it as it is situated of the coast in semi-arid area of Rayalseema region of A.P. The average annual rainfall varies between 550 and 800 mm. The major portion (60%) of the rainfall comes from SW monsoon during the period June to September and a minor amount of rainfall (30%) is due to NE monsoon during the period October to November. Evapotranspiration in this basin is also high. The average rainfall is $615\,mm/year$.

The chief soil types of this basin are red loamy and black cotton soils. The maximum thickness of the black cotton is as much as 9 m in Pennar plain. Red soils are thin as compared with black soil. They range in thickness from few cm to 2.5 m. Alluvium is restricted to the banks of river and streams. It generally comprises sands, gravels, silt and clay having 5–10 m thickness. The confluence of Kunderu with the Pennar is marked by a wide expense of sandy tract. The basin located within the Cuddapah tectonic basin, consists of two groups namely gneisses and granites of Archaean, Cuddapah, and Kurnool super group of Proterozoic age. Quartzites, shales and, limestones of Cuddapah and Kurnool supergroup occupy quite a large area in this basin.[1] Shales (Tradipatri, Cumbum, Auk, and Nandyal) are compact and impervious but are intensively fractured and jointed and therefore develop better groundwater storage conditions. Limestones (Vempalle, Nargis, Koilkuntla, and Nandyal) in this region form a very important group of rocks in the context of groundwater development of this region. Caverneous limestones are the most potential aquifers in this region. The other groups of limestones, which are massive and hard, also yield groundwater through their joints, bedding planes and fracture zones but with lesser quantity. Alluvial deposits along the major and minor river valleys of the basin form potential groundwater reservoir.[2]

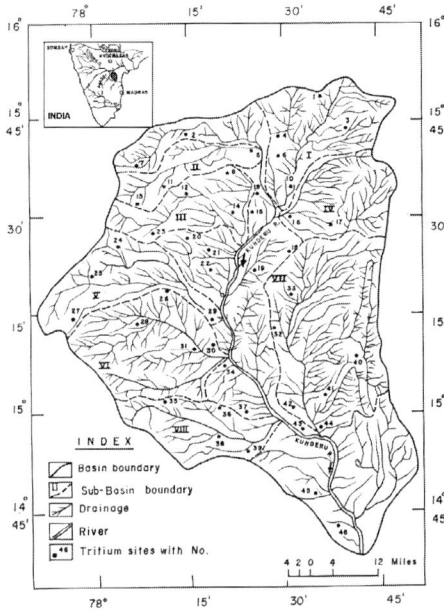

Fig. 1.   Drainage map of Kunderu basin showing tritium injection sites.

3. Recharge Measurements

Artificial tritium injection experiments have been carried out in Kunderu river basin with the aim of calculating average recharge and total input to the shallow groundwater reserves of the basin due to monsoon precipitation. Tritium (H^{3+}), a radioactive isotope of hydrogen, is a soft β-emitter having mass 3 and a half-life of 12.26 years. It exists in the form of water molecules HTO and as such it is an ideal tracer for studying movement of water. The use of tritium as a tracer in recharge measurement involves injection of tritiated water at a certain depth in the soil column and study of the vertically movement that this tracer undergoes during the hydrological cycle. This application of tritium in recharge measurements is based on an assumption of the Piston-Flow model, proposed by Zimmermann and others and Munnich,[3–5] assumes that the percolating soil moisture moves downward in discrete layers and any addition of a fresh layer of moisture at the surface, would push down an equal amount of water immediately below and so on, till the moisture of the last such layer in the unsaturated zone is added to the saturated regime or the water table. The principal steps involved for calculating recharge have been described by earlier researches.[6–9]

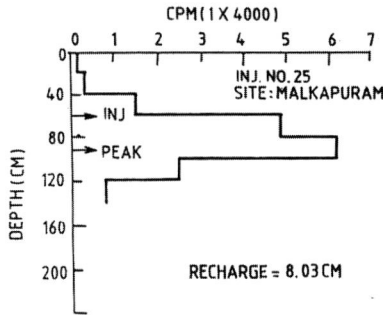

Fig. 2.   Recharge profile at Malkapuram.

46 sites were selected in Kunderu basin and tritium was injected in the month of July before the commencement of monsoon rains. The depth of injection was 80 cm. The sites were distributed uniformly in various soil types covering different geological formations. The location of tritium injection sites in Kunderu basin is shown in Fig. 1. Tritiated water (2.5 ml) having an activity of $10\,\mu C/ml$ was injected at selected depth in a hole made by using drive rod. Vertical soil core profiles were collected from the injected sites with a 20 cm sampling interval after the monsoon cessation. The maximum depth of collection was 250 cm. Moisture content of the samples was determined with torsion balance. The soil samples were subjected to partial vacuum distillation for extracting soil moisture. 4 ml of the distillate was mixed with 10 ml of Instagel (Scintillator cocktail) in low potash glass vials (capacity 20 ml) and the tritium activity of the sample was counted by Scintillation spectrometer. The recharge in the case of each site was calculated by determining the peak/center of gravity of the tritium versus depth profiles. The mean displacement of the tracer is calculated from the depth of injection to the peak/center of gravity. The net recharge was taken as the length of an imaginary water column over a square centimeter in the soil profile interval between the depth of injection and the depth of peak/center of gravity. Recharge was computed using tracer displacement and moisture content data. Figure 2 shows typical tritium profile of Kunderu river basin at Malkapuram.

4. Results and Discussions

The recharge values were found to vary from 26 to 94 mm. The arithmetic mean of all the recharge values is 29 mm with a standard deviation of 24 mm.

The Kunderu basin is further divided in to eight subbasins namely Galeru (I), Madieru (II), Kollamvagu (III), Palleru (IV) Jerruru (V), Palieru Peddavanka (VI), Vakileru (VII), and Tigaleru (VIII) as shown in Fig. 1. The average recharge values obtained from injected sites in these subbasins are 20.4, 59.3, 17.5, 44.1, 49.3, 38, 12.3, and 37.1 mm, respectively. Few sites in the basin show negative recharge. These are sites where practically negligible or no recharge takes place (discharge area).

Raju and others[10] have indicated that greater than 5% slope cannot hold surface water for sufficient time to allow the percolation and so the area above 5% slope can be neglected for calculating the effective recharge area in a basin. Using this criterion, the effective infiltration area of Kunderu basin was calculated as 7,241 km^2. In total 211 million cubic meters (MCM) in Kunderu basin has been recharged to groundwater and was added to near surface aquifers due to monsoons. This is equivalent to average net recharge of 0.029 MCM over an sq. km area.

The recharge estimates at Kunderu river basin from 46 sites range from -25.8 to 94.3 mm with a mean value of 29 mm. This is 4.7% of the monsoon rainfall of 615 mm. The measured recharge value shows a wide scatter, ranging from 29 to 179 mm. This is due to variation in pattern and magnitude of rainfall, change in soil type and its heterogeneity. The total annual input to groundwater reserves of Kunderu river basin due to monsoons rains over an effective infiltration area of 7,241 sq. km was 211 MCM. The reproducibility of recharge measurements obtained by tritium injection method has been found to be +10% from the mean. The sites found, through tritium injection studies, to be giving high recharge values, can be considered favorable for implementation of schemes of artificial recharge.

Acknowledgments

The authors wish to thank to Dr. V.P. Dimri, Director, NGRI, Hyderabad for his kind support and his permission to publish this paper.

References

1. S. Balakrishna, B. Venkatnarayana and M. N. Rao, *IIPG*, Hyderabad, 1979.
2. I. Radhakrishna, *IIPG*, Hyderabad, 1982.
3. U. Zimmermann, D. Ehhalt and D. O. Mummich, *Isotopes in Hydrology*, IAEA, Vienna, 1967a, pp. 567–586.
4. U. Zimmermann, K. O. Munnich and W. Roether, *AGU Geophysical Monograph* **11** (1967b) 28–36.

5. K. O. Munnich, *Guide Book on Nuclear Techniques in Hydrology* (IAEA, Vienna, 1968), pp. 112–117.
6. R. Rangarajan and R. N. Athavale, *Journal of Hydrology* **234** (2000) 38–53.
7. R. Chand, S. Chandra, V. A. Rao, V. S. Singh and S. C. Jain, *Journal of Hydrology* **299** (2004) 67–83.
8. R. N. Athavale, R. Chand and R. Rangarajan, *Hydrological Sciences-Journal* **28**, 4 (1983) 12.
9. R. Chand, G. K. Hodlur, M. R. Prakash, N. C. Mondal and V. S. Singh, *Curr. Sci.* **88**, 5 (2005) 10.
10. K. C. C. Raju, M. Karemuddin and P. P. Rao, *Miscellaneous Publication*, No. 47, GSI, Government of India, (1979), p. 39.

SIMULATION OF SUBSURFACE WATER, NUTRIENTS, AND CONTAMINATION DISCHARGE TO COAST

A. GHOSH BOBBA

National Water Research Institute, Environment Canada
Burlington, ON, Canada L7R 4A6
ghosh.bobba@ec.gc.ca

Ecosystems of the coastlines are receiving extraordinary amounts of nutrients as a consequence of human activities such as fertilizers, industrial emissions to the atmosphere, and disposal of waste water in coastal watersheds. The loadings of nitrogen and phosphorous to coastal aquatic environments even exceed those to fertilized agro-ecosystem. Increased nutrient loading from anthropogenic sources is pervasive and function of shallow coastal ecosystems during coming decades.

During the last decade, it has become apparent that subsurface water flow and transport of nutrients into shallow coastal water are far more significant and widespread than had been realized. The importance of subsurface water is not so much because of the magnitude of flow rates, but rather because of the high nutrient concentrations in subsurface water compared to those in receiving sea/lake water. Although highly variable, the nutrient content of subsurface water discharging onto coastal water may be up to five orders of magnitude larger than concentrations in receiving sea/lake water.

Subsurface water discharge has shown to be a source of nitrogen, typically as nitrate, in shallow sediments of lakes and coastlines. The response of primary producers to nutrient loading within an estuary must be dependent on the balance between increased growth due to elevated inputs of the limiting nutrient and losses related to the flushing rate, as per the model. This is the point of the well-documented relation between phosphorous loading and phytoplankton chlorophyll.

In this paper, some Canadian and Indian examples show the roles of geo-hydrology and hydro-geochemical processes are investigated by the application of numerical models for water quantity and quality modeling of receiving waters. This model also predicts the movements of contaminants in subsurface and surface waters and coastal sediments.

1. Introduction

Subsurface water discharge from watersheds to coastal and estuarine waters has been a topic of theoretical and practical interest for at least a century.[1] Most work has stressed controls on saltwater intrusion to freshwater aquifers and it has been only recently that field measurements of groundwater discharge have shown the importance of subsurface flow on water and nutrient budgets in coast and estuaries. Groundwater discharge has shown to be a source of nitrogen, typically as nitrate, in shallow sediments of lakes and coastlines. For example, upland aquifers contribute to direct groundwater discharge greater than 20% of the freshwater and 75% of the nitrogen that enters the Great South Bay, New York.[2] Excess nitrogen in groundwater derived from sewage and fertilizer has drastically affected water quality in other estuaries and coastal lagoons as well.[3]

Ecosystem of the world's coastlines are receiving extraordinary amounts of nutrients as a consequences of human activities such as fertilizers, industrial emissions to the atmosphere, and disposal of waste water in watersheds adjoining coastal waters. The loadings of nitrogen and phosphorous to coastal aquatic environments even exceed those to fertilized agro-ecosystem. Increased nutrient loading from anthropogenic sources is pervasive and function of shallow coastal ecosystems during coming decades.

Although increased nutrient loading by precipitation has been documented, most research has focused on deeper estuaries in which flow from rivers and streams dominates water budgets and contributes the major of nutrients. Rivers and direct precipitation, however, are not the sole source of freshwater — borne nutrients to coastal environments. Even in places without rivers, salinity is often depleted in coastal waters due to groundwater input. Groundwater flow is especially important where underlying coastal sediments are coarse, unconsolidated sands of glacial or marine origin. In such situations flow of groundwater may be the major source of nutrients to coastal waters.

In unconsolidated sediments, groundwater moves through the watershed shoreward in paths that have downward vertical as well as horizontal vector. Downward flow is caused by additional water infiltrating along the path of the water. Freshwater eventually moves close enough to shore to meet the denser saltwater that saturates interstitial space in sediments beneath the sea. The presence of seawater in the pore space acts together with lower head pressures in the near shore zone compared with offshore to deflect the path of fresh groundwater sharply upward.[1] As a result, most of the groundwater flow occurs very near the shore.

During the last decade, it has become apparent that groundwater flow and transport of nutrients into shallow coastal water are far more significant and widespread than had been realized. The importance of groundwater is not so much because of the magnitude of flow rates, but rather because of the high nutrient concentrations in groundwater compared to those in receiving coastal water. Although highly variable, the nutrient content of groundwater discharging onto coastal water may be up to five orders of magnitude larger than concentrations in receiving seawater.

2. Consequences of Groundwater Nutrient Transport

There is enormous freshwater literature that deals with the consequences of eutrophication. The response of primary producers to nutrient loading within an estuary must be dependent on the balance between increased growth due to elevated inputs of the limiting nutrient and losses related to the flushing rate. This is the point of the well-documented relation between phosphorous loading and phytoplankton chlorophyll. Additional evidence, based on correlations and inference from N:P values, on enrichments within small containers, and on whole-system enrichments in freshwaters, is consistent with the phosphorus loading results, and shows that in freshwater phosphorus limits potential primary production.

In coastal systems, the evidence of the role of nitrogen in limiting phytoplankton growth is based on inferences from loading calculations, on ambient nutrient concentrations, and more importantly, on nutrient enrichment experiments.

Groundwater entering the coastal areas flows from watersheds where urbanization has taken place. Human activities such as disposal of wastewater via septic tanks, and use of fertilizers increase nutrient concentrations in groundwater. We presently lack of data groundwater along coastal areas, but can make preliminary comparisons using data from nearby locations, where watersheds include some urbanized areas. This paper focuses on application of numerical model to predict subsurface water and pollution discharge to coast.

3. Subsurface Hydrology of a Coastal Watershed

The two main sources of pollutants are point sources (Table 1) and nonpoint sources (Table 2). Pollutants from the two sources may be released continuously or at discrete intervals (Fig. 1). Point sources of pollution can be geometrically defined and the dimensions are amenable to mathematical analysis in assessing pollution loads and rates of discharge determined. Distributed sources of pollutants are much more widespread and can rarely be geometrically defined as precisely as a point source. Hence, it is more

Table 1.   Point sources of pollution in a coastal watershed.

Type of pollution	Examples
Sewage disposal systems	Sewage lagoons, septic systems, cesspools, barnyards/feed lots
Surface waste disposal sites	Landfills, garbage dumps, and surface waste dumps
Underground waste disposal sites	Storage tanks (low, medium, high level wastes)
Spills, washings, and intrusions	Oil, gas, waste spills: auto workshop washings, Research laboratory washings, seawater or saltwater intrusions
Mining sources	Acid mine drainage: mine waste dumps, and seepages gas explosions
Natural mineral/ore deposits	Saline springs, hot spring waters, anhydrite, pyrite deposits, etc.

Table 2.   Non-point sources of pollution in a coastal watershed.

Source	Examples
Agriculture	Cropland, irrigated land, woodland, and feed lots
Silviculture	Growing stock, logging, and road building
Construction	Urban development, and highway construction
Mining	Surface, and underground
Utility maintenance	Highways, streets, and deicing
Urban runoff	Floods and snowmelt

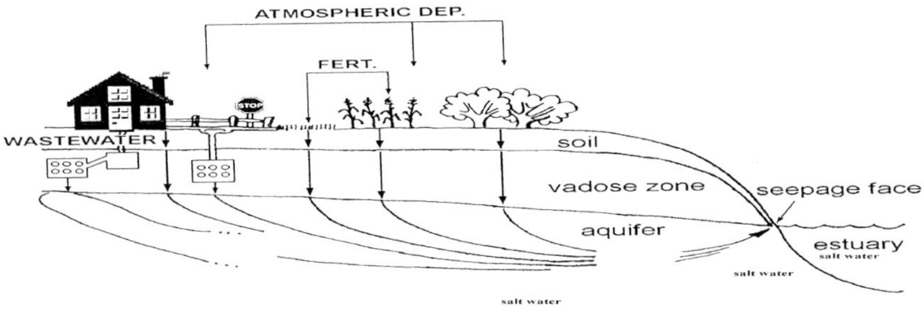

Fig. 1.   Different sources of pollutants in a coastal watershed.

difficult to subject the input/output source to precise mathematical analysis. Rather a measured and intelligent assumption of the affected area is made for use in modelling and analysis. In heavily polluted areas, both point sources and distributed sources may occur together or may be independent of one another.

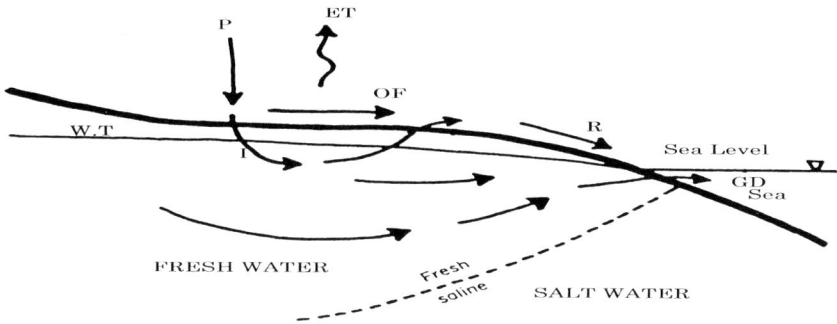

Fig. 2. Hydrological pathways in a coastal watershed.

A watershed is a topographically defined as an area that water enters through precipitation and leaves as evapotranspiration, surface runoff, and subsurface water discharge. In the case of a coastal watershed, runoff and subsurface water discharge enter the sea (Figs. 1 and 2). Rainfall and potential rate of evapotranspiration are determined by climate. Actual evapotranspiration is limited by the climatically controlled potential rate, vegetation and the wetness of the soil. Soil properties, topography, and the history of rainfall and evapotranspiration determine runoff and subsurface water discharge. For the purpose of this discussion, surface runoff includes direct runoff which occurs as stream flow immediately following a rainstorm, and drainage of subsurface water into creeks which accounts for stream flow between streams. The amount of direct runoff generated by a storm depends on the amount of rainfall and on the moisture condition of the soil. In general, more runoff occurs when the soil is initially wet. The subsurface water discharge can be calculated if rainfall, evapotranspiration, runoff, and the change in the amount of water stored on the watershed are known.

A link between rising sea level and changes in the water balance is suggested by the general description of the hydraulics of subsurface water discharge at the coast. Fresh subsurface water rides up over denser saltwater in the subsurface system on its way to the sea (Fig. 2), and subsurface water discharge is focused into a narrow zone that overlaps with the intertidal zone.

The width of the zone of subsurface water discharge measured perpendicular to the coast is indirectly proportional to the discharge rate. The shape of the water table and the depth to the fresh/saline interface are controlled by the difference in density between freshwater and saltwater, the rate of freshwater discharge and the hydraulic properties of the subsurface system. The elevation of the water table is controlled by mean sea level through hydrostatic equilibrium at the shore.

Because pollutants are transported in large part by the bulk motion of subsurface water, the parameters of subsurface water flow are of major importance in the understanding of pollution processes. The various aspects of the subsurface water environments, as well as stratigraphic factors that control or could influence subsurface water motion are also of major consideration. The subsurface hydrology environment is shown schematically in Fig. 2. It consists mainly of saturated and unsaturated zones. The unsaturated zone occurs above the capillary fringe, where the soil pores are partially saturated with water. This zone is important in waste management because in most cases, it is the burial zones for wastes. Consequently, a thick unsaturated zone may sometimes be preferred for waste disposal since it would take a much longer time for pollutants to reach the water table. In the saturated zone, the pores are saturated with water. When this zone is capable of transmitting significant quantities of water for economic use it is referred to as an aquifer. In most field situations, two or more aquifers occur, separated by impermeable strata or aquitard. In the situation illustrated in Fig. 2, the upper or unconfined aquifer is much more prone to pollution than the lower confined aquifer.

3.1. Numerical simulation of subsurface and contamination discharge to coast

Bobba and Singh[4] presented a detailed description of available model characteristics and recent trends in subsurface water flow and contamination transport modeling. Flow models are used to determine the quantitative aspects of subsurface water motion, such as direction, rate, changes in water table or potentiometric head, stream-aquifer interaction, etc. Transient 2D models for simulating groundwater flow are widely available.

One common goal of these models is to predict and characterize the movement of the transition zone in the aquifer where freshwater and saltwater meet in coast. Another purpose of modeling is to predict the degree and extent of mixing that occurs in this transition region. In this way, models allow problems to be defined before they actually occur. The details of the models presented earlier by Bobba,[5] Bobba and Singh,[4] Bobba et al.[6]

In this paper, a finite element model, SUTRA[7] is applied to three different types of cases. This model simulates water movement and the transport of either dissolved substances or energy in the subsurface system. The model can be applied areally or in cross section. It uses a 2D, combined finite-element and integrated-finite difference method to approximate the equations that describe the two interdependent processes being simulated. When used to simulate saltwater movement in the subsurface system in cross section, the two interdependent processes are the density-dependent saturated subsurface-water flow and the transport of dissolved solids in the subsurface water. Either local — or regional — scale sections having dispersed or relatively sharp transition zones between saltwater and fresh-water may be simulated. The results of numerical simulation of saltwater movement show distributions of fluid pressures and dissolved-solid concentrations as they vary with time and also show the magnitude and direction of fluid velocities as they vary with time. Almost subsurface properties that are entered into the model may vary in value throughout the simulated section. Sources and boundary conditions may vary with time. The finite-element method using quadrilateral elements allows the simulation of irregular areas with irregular mesh spacing. The model has been applied to real field data and observed to give favorable are explained as follows.[5-8]

3.2. *Case study 1: Lambton County, Ontario, Canada*

Lambton County is located in the province of Ontario, Canada (Fig. 3). The study region occupies most of Lambton County is located along the Saint

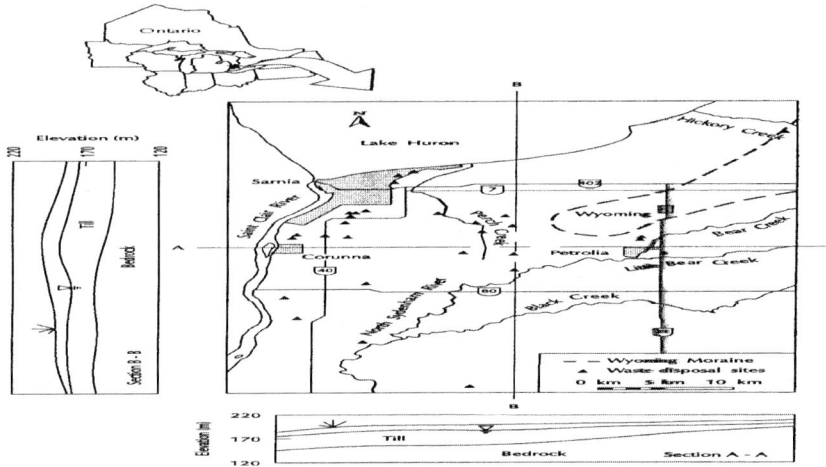

Fig. 3. Location of Lambton County, where Sections A-A and B-B show the surface geology and hydrogeology of region in East-West and North-South cross sections (Piggott, Bobba and Novakowski, 1996).

Clair River between Lakes Huron and Saint Clair. The region is boundary to Lake Huron and to the west by the Saint Clair River. Numerous chemical and petrochemical companies operate in this region and the wastes generated by these facilities have historically been managed using near surface burial and deep well injection. Sporadic discharge of formation fluids, and possibly industrial waste, to the surface has caused concern over the contamination of Lake Huron, the Saint Clair River, and the regional aquifer that serves as a water supply for rural Lambton County. Migration of wastes injected at depth to the near surface via discontinuities in the confining strata and abandoned deep wells is a plausible mechanism for the contamination of surface and subsurface waters. The numerical models are used to determine rates and directions of groundwater flow for contamination transport studies, predict the impact of varying climatic conditions and groundwater consumption, guide sampling programs for natural and anthropogenic contamination, and determine groundwater discharge to surface water bodies. The geology and hydrogeology of the site are described earlier.[1,8] The groundwater is obtained from sand and gravel deposits at depths below 60 m. The aquifer is called as Freshwater Aquifer that is floating on top of high saline water due to density difference. Flow in this aquifer is likely be strongly influenced by precipitation events and results in either discharge to shallow ditches, creeks, and rivers of recharge to the deeper formations.

The application of the numerical explained earlier.[8] Figure 4 shows the computed hydraulic head map which best approximates the observed water

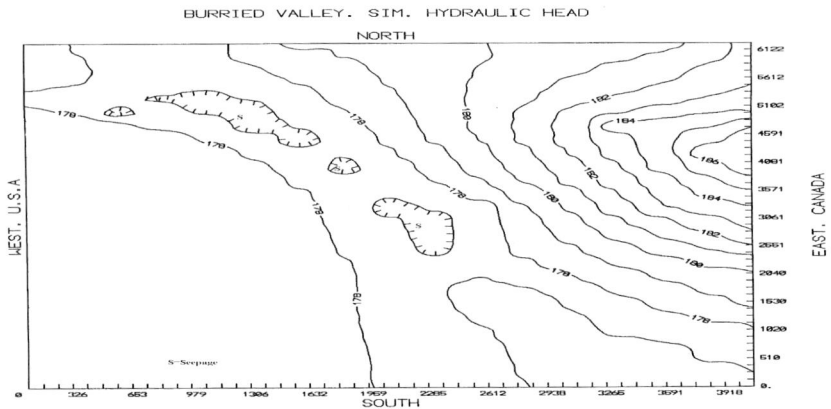

Fig. 4.   Simulated hydraulic head map of the buried valley, and groundwater seepage locations in Saint Clair River, Canada, and USA boundary.[1,8]

table. The model results indicate that virtually all of the water flowing through the aquifer discharges to the St. Clair River and Lake Huron. The water level contours in the bedrock valley indicate a generally westward groundwater flow system. Since hydraulic gradient varies across the bedrock valley, the rates of groundwater flow also vary. Calculations indicate that groundwater flux into and out of the valley are about 0.9 and $0.5\,m^3/s$, respectively, for a 100 m strip of aquifer. The amount which flows westward out of the $(0.5\,m^3/s)$ probably discharges to the St. Clair River. Discharge of groundwater from the Canadian side of the Freshwater Aquifer is calculated to be between 0.45 and $0.50\,m^3/s$ for that portion of the river between Lake Huron and Stag Island.

3.3. *Case study 2: Port Granby radioactive disposal site*

Canada has a uranium refinery, at Port Hope, Ontario. The waste from refinery was disposed at Port Granby waste management site located on the north shore of Lake Ontario (Fig. 5). In recognition of concern over the possible contamination of surface lake waters, the concentrations of radium and uranium were measured in water samples collected of Lake Ontario coastal zone near waste site. These data showed that the leachate infiltrating and seeping to coastal zone of Lake Ontario. The plume, moving parallel to the shoreline in the direction of the prevailing wind direction (Fig. 5). The finite element model applied to calculate hydraulic head and contamination discharge to lakeshore. The predicted Ra-226 contamination

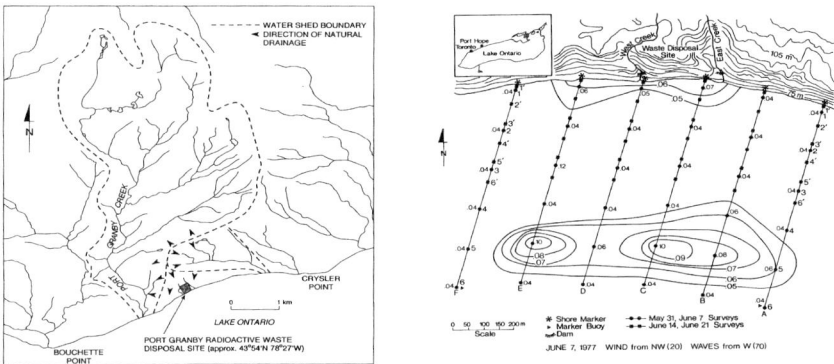

Fig. 5. Location of waste disposal site and Ra-226 concentration in Lake Ontario, Canada.

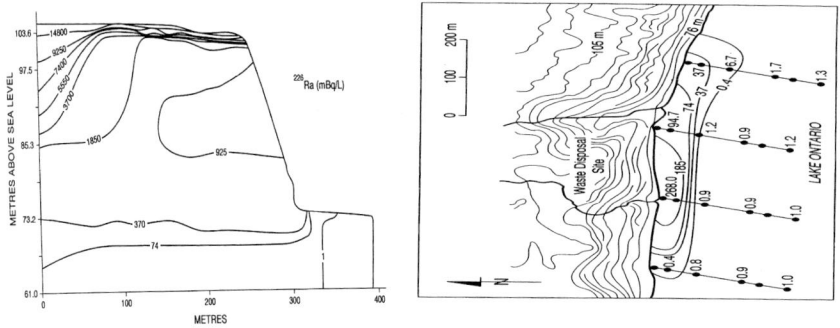

Fig. 6.   Computed Ra-226 concentration in waste site, beach, and observed concentration in coast.[9]

concentration from waste disposal site to beach is shown in Fig. 6. The groundwater discharge from the waste disposal site between the east and west creeks is computed by applying Darcy's law.

The groundwater flow to Lake Ontario from the site is $2.5 \times 10^5\,\mathrm{m^3/year}$. This annual volume of water is about 3%–4% of the total solid volume of the site, assuming a depth of about 35 m. If the ^{226}Ra activity in the shore piezometers averages about $100\,\mathrm{Bq/m^3}$, about $2.5 \times 10^7\,\mathrm{Bq/year}$ is carried into the lake. But the total amount of ^{226}Ra which has been disposed of at the site is $2.3 \times 10^{13}\,\mathrm{Bq}$, so that only about 10^{-6} of this is lost from the site per year. A comparable calculation for uranium indicates that about 25 kg of this element reaches the lake. The details of the application of model and interpretation of results were presented earlier.[9]

3.4. Case study 3: Godavari delta, India

The Godavari delta is located in East Coast of India (Fig. 7). The details of geology and environmental problems have been explained earlier.[10] The Godavari delta lies between the sea level and 12 m contour. The delta has a projection of about 35 km into the sea from the adjoining coast.

The Godavari delta consists of alluvial plain. It has a very gentle land slope of about 1 m per km. The coastal line along the study area measures to about 40 km and the general elevation varies from about 2 m near the sea to about 13 m at the upper reach. Texturally, a major part of the study area consists of sandy loams and sandy clay loams. The silty soils, which are very deep, medium textured with fine loamy soils is located all along the Godavari River as a recent river deposits. The very deep, coarse textured

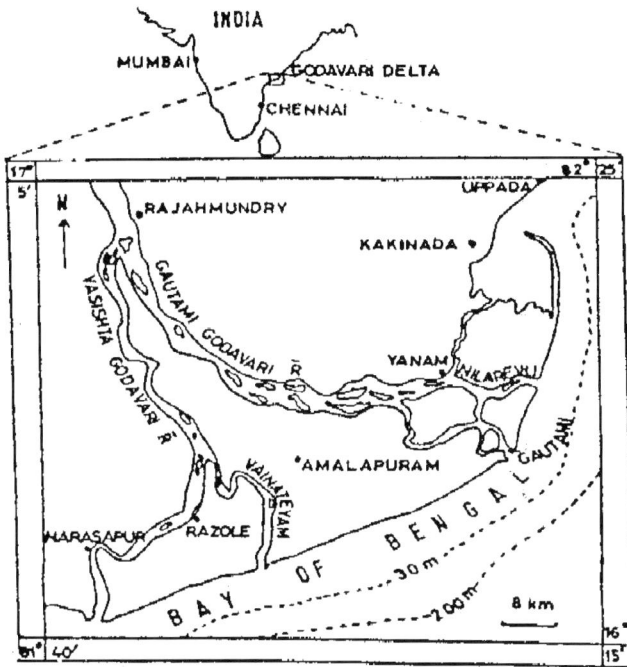

Fig. 7.   Location of the Godavari delta.

soils with sandy subsoils representing the coastal sand are also found along the sea.

The delta has a network of canal systems and increased use of new crops along with chemical fertilizers and pesticides have brought about rapid growth in the agricultural output. The crops irrigated with river water throughout the year except between the last week of April and the second week of June. As ample surface water is made available to irrigate the delta, there has been no effort to use groundwater. Potassium fertilizers are extensively used for increasing the crop yield. The variation in potassium concentrations in groundwater in the study area is shown in Fig. 8. The peak values are generally observed in November. Higher concentrations are found due to the application of potash fertilizer besides contribution from soils. The temporal variation of chloride and bicarbonate from groundwater samples is shown in Fig. 9.

The ratio of chloride/bicarbonate + carbonate can be used as criteria to evaluate seawater intrusion. Chloride is the dominant ion in seawater and

Concentration of K in PPM

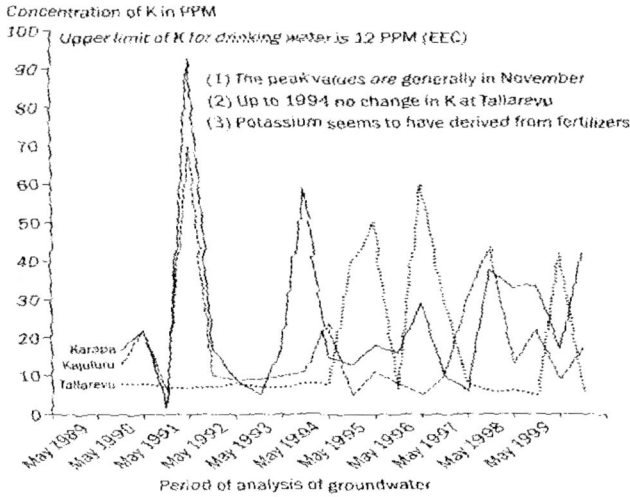

Fig. 8.   Variation of potassium concentration in groundwater in the Godavari Delta.[12]

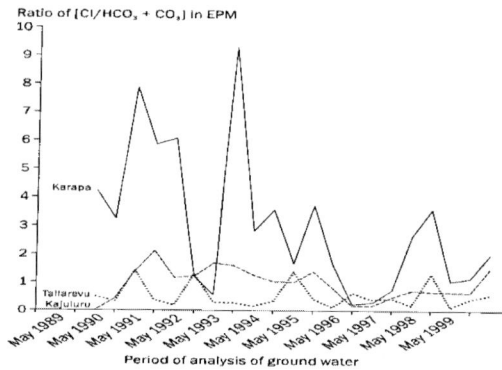

Fig. 9.   Variation of [Cl/HCO$_3$+CO$_3$] ratio in groundwater in the Godavari Delta.[12]

it is only available in small quantities in groundwater while bicarbonate, which is available in large quantities in groundwater, occurs only in very small quantities in seawater. The details of the model and application to the Godavari delta basin have been explained earlier by Bobba.[10,11] The prediction of the water table depth due to irrigation and saltwater intrusion reported earlier.[10,11] Figures 10(b), and 11(b) depict the influence of non-irrigation and irrigation (rainy) seasons. During the irrigation season periods, the water table is raised due to irrigation (Fig. 11(b)), water recharged

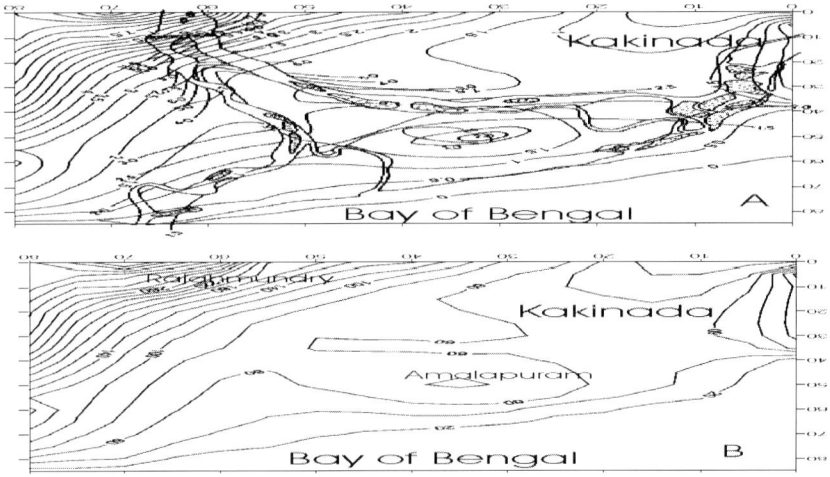

Fig. 10.   Simulated hydraulic heads (A) and freshwater depth (B) of Godavari delta in non-irrigation months.[10]

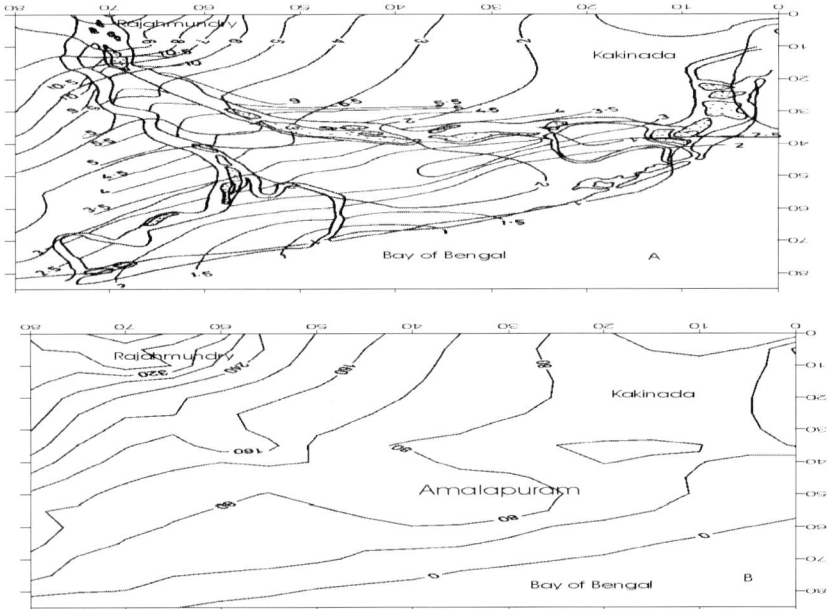

Fig. 11.   Simulated hydraulic heads (A) and freshwater depth (B) of Godavari delta in irrigation season months.[10]

to subsurface. The distance between surface and water table in the coastal area is very small, and the material is generally composed of sands, which do not retain significant amounts of moisture under unsaturated conditions. Hence, the irrigated water directly recharges to the unsaturated zone. The distance between the water table and surface is at a minimum in the central portion of the delta. It has been observed that areas of minimum depth from the ground level to the water table have high freshwater potential whereas lowering of the water table (Fig. 10(b)) from the ground surface reduces the freshwater potential substantially. The water table elevation varies from 0.5 to 1 m from MSL and decreases gradually toward the coast.

Figure 12 shows the results in different environmental conditions due to heavy and long rainy season and drought conditions due to high temperature and less rainfall (climate change). Higher water table conditions are observed due to heavy rain and irrigated water is recharged to the

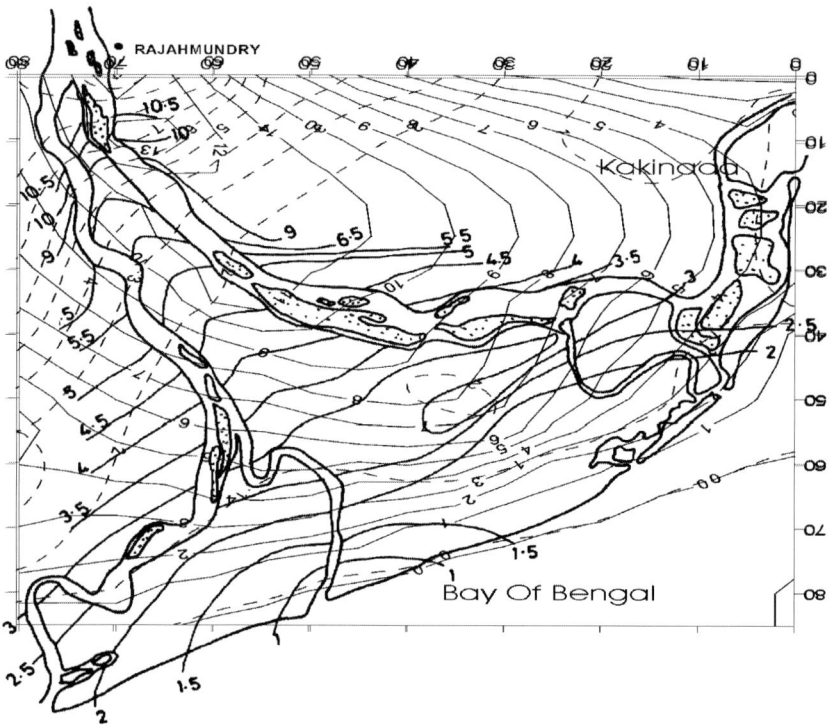

Fig. 12.   Simulated hydraulic heads of Godavari delta in different seasons (solid line in heavy rainy season, - - - drought conditions).[10]

aquifer. The saltwater was flushed out or stopped saltwater intrusion to the aquifer. However, if the severe drought conditions (higher temperature, lesser rainfall) occur in the delta, the water table is reduced due to higher evapotranspiration and over pumping the ground water for irrigation and domestic purposes. The saltwater intruded to the aquifer and freshwater thickness reduced in the delta.

4. Summary and Conclusions

Mathematical and numerical models are useful to estimate subsurface water and contamination discharge to coastal areas from point and non-point source areas. Applications of two types of numerical models were presented in this paper. As an example, a finite element model, considering open boundary conditions for coasts and a sharp interface between freshwater and saltwater was applied under steady-state conditions to the phreatic aquifer for freshwater surplus and deficits at the coastline due to El Nino effect. When recharges of saltwater occur at the coastline, essentially of freshwater deficits, a hypothesis of mixing for the freshwater-saltwater transition zone allows the model to calculate the resulting seawater intrusion in the aquifer. Hence, an adequate treatment and interpretation of the hydrogeological data that are available for a coastal aquifer are of main concern in satisfactorily applying the numerical model.

The results of the steady-state simulations showed reasonable calculations of the water table levels and the freshwater and saltwater thicknesses, as well as, the extent of the interface and seawater intrusion into the aquifer for the total discharges or recharges along the coastline. As a result of the present hydrogeological simulations on the subsurface system, a considerable advance in seawater intrusion would be expected in the coastal watershed if the sea level rises due to climate change and El Nino effects.

References

1. A. G. Bobba, Mathematical models for saltwater intrusion in coastal aquifers — literature review, *Water Resources Management* **7** (1993a) 3–37.
2. H. J. Bokuniewicz, Groundwater seepage into Great South Bay, New York, *Estuarine and Coastal Marine Science* **10** (1980) 437–444.
3. I. Valiela, K. Foreman, M. LaMontagne, D. Hersh, J. Costa, Peckol DeMeo-Anderson, C. D'Avanzo, M. Babione, C. Sham, J. Brawley and K. Lajyha, Coupling of watersheds and coastal waters: Sources, and consequences of nutrient enrichment in Waquoit Bay, Massachusetts, *Estuaries* **15** (1992) 443–457.

4. A. G. Bobba and V. P. Singh, Groundwater contamination modelling, *Environmental Hydrology*, ed. V. P. Singh (Kluwer Academic Publishers, Dordrecht, 1995), Chap. 8 pp. 225–319.

5. A. G. Bobba, Application of numerical model to predict freshwater depth in islands due to climate change effect: Agati Island, India, *Journal of Environmental Geology* **6** (1998) 1–13.

6. A. G. Bobba, V. P. Singh and L. Bengtsson, Application of environmental models to different hydrological systems, *Journal of Ecological Modelling* **125** (2000) 15–49.

7. C. I. Voss, SUTRA — A finite element simulation model for saturated — unsaturated fluid density dependent groundwater flow with energy transport or chemically reactive single species solute transport, USGS Water Resource, Investigation Report 84-4269 (1984).

8. A. G. Bobba, Field validation of "SUTRA" groundwater flow model to Lambton County, Ontario, Canada, *Water Resources Management* **7** (1993b) 289–310.

9. A. G. Bobba and S. R. Joshi, Groundwater transport of radium-226 and uranium from Port Granby Waste management site to Lake Ontario, *Nuclear and Chemical Waste Management* **8** (1988) 199–209.

10. A. G. Bobba, Numerical modelling of saltwater intrusion due to human activities and sea level change in the Godavari delta, *Hydrological Sciences Journal* **47**(S) (2002) S67–S80.

11. A. G. Bobba, Numerical simulation of saltwater intrusion into coastal basin of Indian sub-continent due to anthropogenic effects. ICIWRM — 2000, *Proceedings of International Conference on Integrated Water Resources Management for Sustainable Development* (National Institute of Hydrology, Roorkee, India 2000), Vol. 1 pp. 323–340.

12. A. G. Chachadi and L. Teresa, Health of the groundwater regime in a coastal delta of East Godavari, Andhra Pradesh, Coastin, *A Coastal Policy Research Newsletter* (TERI, 9607/13, Multani Dhanda, Paharganj, New Delhi, India, 2002), pp. 5–8.

MODELING OF PHYSICAL AND GEOCHEMICAL BEHAVIORS OF SALTWATER IN A COASTAL AQUIFER

KEIGO AKAGI*,‡, TOSAO HOSOKAWA†, YOSHINARI HIROSHIRO*

and KENJI JINNO*

*Institute of Environmental System, Kyushu University
6-10-1, Hakozaki, Higasiku, Fukuoka, Japan
‡hyd22@civil.kyushu-u.ac.jp

†Department of Urban Infrastructure, Kyushu Sngyo University
2-3-1, Matsukadai, Higashiku, Fukuoka, Japan

Groundwater quality is affected by the various reactions such as precipitation of dissolved ions, solid–liquid phase interaction, and the bacteria-mediated reduction–oxidation process. The geochemical properties in coastal aquifers are unique and important for the management of groundwater environment. Although the studies analyzing the hydrological behaviors have been intensively done up to now, a study of coupling both groundwater flow and geochemical analysis is limited. In this study, the one-dimensional multicomponent solute transport model which accounts for the cation exchange reaction and the reduction reaction induced by anaerobic degradation is studied. It is found out both through experiment and numerical simulation that the cation exchange process is significant at the frontal part of the seawater infiltration, while reducing process by the bacteria mediation takes places at the entire soil column except for the top of the soil where dissolved oxygen infiltrates. Saltwater intrusion into a coastal aquifer is a traditional but still important for the hydro-geologists and civil engineers even at present time. Contamination of salt is a trouble for various freshwater uses. Therefore, much attention has been put on the movement and potential contamination. Meanwhile, the geophysical study is solely carried out. However, the consistent and accurate conclusion will be hardly obtained due to the complicated and limited information on the aquifer properties. In order to get a confident conclusion that can be agreed by the researchers in various fields, an interdisciplinary approach anticipating both hydrological and geochemical processes is indispensable.

1. Introduction

Although several computer codes are available at present to calculate the transport of multi-chemical species in aquifers, the properties of response of entity species are very complicated and difficult to appropriately set the necessary soil geo-chemical parameters when many chemical elements are involved. Specifically, bacteria-mediated reducing process in anaerobic condition requires the parameters representing the bacteria growth and degradation.[1] Kinzelbach *et al.*[2] developed the aerobic-anaerobic mass transport model in which oxygen, nitrate, organic carbon and bacteria growth in liquid, solid, and biophases are included. Lensing *et al.*[3] included the effect of inter-phases mass transport related to the bacteria-mediated biochemical reactions. Schäfer *et al.*[4,5] examined the effect of decrease of organic carbons consumed by bacteria on the biochemical mass transport. The inclusions of sulfur reducing and methane yielding processes were considered important in the modeling of the extremely reduced aquifer conditions. The authors developed the algorithm of ion exchange reactions for the major cation such as Ca^{2+}, Mg^{2+}, K^+, Na^+, and the dissolved Mn^{2+} and Fe^{2+}, which are yielded by the biochemical reactions by bacteria, considering the possible contribution on the cation exchange process.[6,7] This effect should be significant for the situation in a coastal aquifer where the intruded seawater region resides for a long time compared to the freshwater above the mixing zone between the fresh and seawater. For example, Snyder *et al.*[8] reported that the reduced environment of Mn^{2+} and Fe^{2+} plays an important role in determining the geochemical properties in a coastal aquifer, where the bacteria mediated biochemical reactions would take place.

In the present paper, objecting the ion exchange processes by six cations such as Na^+, K^+, Ca^{2+}, Mg^{2+}, Mn^{2+}, and Fe^{2+}, the consumptions of dissolved oxygen (DO) and nitrate caused by aerobic bacteria and the reduction of manganese oxidize and iron hydroxide were studied through the laboratory experiment infiltrating seawater into the alluvial soil. The numerical simulation model including the above mentioned chemical species in the bio-, liquid-, and solid phases were also evaluated. By conducting a sensitivity analysis for the model parameters, several significant properties in the mass transport were obtained. This study, particularly, focuses on the iron behavior, because iron oxide abundantly exists in nature, iron is common redox element and has much effect on redox mechanism.

2. Mult-icomponent Solute Transport Model

Figure 1 illustrates the conceptual model of the reactions considered in this research. The reactions to be discussed herein are the cation exchange, bacteria-mediated biochemical processes as well as the convection and dispersion. The bacteria-mediated process is assumed to take place in the biophase. Arrows in the figure depict the possible mass transport between the different phases.

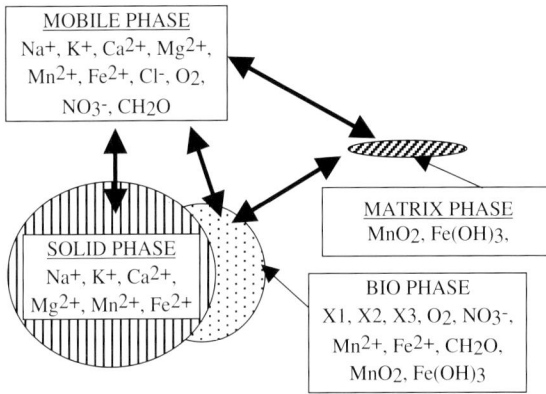

Fig. 1.   Conceptual model.

The one-dimensional mass transport equation is represented by Eq. (1) when the y-coordinate is taken along the depth of the experimental apparatus.

$$\frac{\partial [C_i]}{\partial t} + v'\frac{\partial [C_i]}{\partial y} = \frac{1}{\theta_{\mathrm{w}}}\frac{\partial}{\partial y}\left(\theta_{\mathrm{w}} D_{\mathrm{L}}\frac{\partial [C_i]}{\partial y}\right) + S_i, \qquad (1)$$

where θ_{w} is the water content, v' the pore velocity, and D_{L} is the longitudinal dispersion coefficient that is calculated by the product of the longitudinal dispersion length and the pore velocity.

Following equations represent the Fe^{2+} concentration change in each phase.

Mobile phase

$$Fe^{2+}: \quad \frac{\mathrm{d}[Fe^{2+}]_{\mathrm{mob}}}{\mathrm{d}t} = \frac{\partial [Fe^{2+}]_{\mathrm{mob}}}{\partial t} + v'\frac{\partial [Fe^{2+}]_{\mathrm{mob}}}{\partial y}$$

$$= \frac{1}{\theta_{\mathrm{w}}}\frac{\partial}{\partial y}\left(\theta_{\mathrm{w}} D_{\mathrm{L}}\frac{\partial [Fe^{2+}]_{\mathrm{mob}}}{\partial y}\right) + \frac{T_{\mathrm{b}}(1-n)}{a}$$

$$\cdot \frac{\theta_{\mathrm{bio}}\sqrt{D_{\mathrm{L}}}}{\theta_{\mathrm{bio}} + \theta_{\mathrm{w}}}([Fe^{2+}]_{\mathrm{bio}} - [Fe^{2+}]_{\mathrm{mob}}) + S3_{\mathrm{Fe}}, \quad (2)$$

Solid phase

$$Fe^{2+}: \quad \frac{\partial}{\partial t}(\theta_{\mathrm{w}}[Fe^{2+}]_{\mathrm{im}}) = -\theta_{\mathrm{w}} S3_{\mathrm{Fe}} \qquad (3)$$

$$[Fe^{2+}]_{\mathrm{im}} = \frac{(1-n)\rho_{\mathrm{s}}}{\theta_{\mathrm{w}}}\bar{m}_{\mathrm{Fe}}. \qquad (4)$$

Biophase

Fe^{2+}: $\dfrac{\partial}{\partial t}(\theta_{bio}[Fe^{2+}]_{bio}) = \dfrac{1}{P_{Fe^{2+}}}\left[\dfrac{\partial\theta_{bio}X3}{\partial t}\right]_{grow} - \dfrac{T_b(1-n)}{a}$

$$\cdot \dfrac{\theta_{bio}\theta_w\sqrt{D_L}}{\theta_{bio}+\theta_w}([Fe^{2+}]_{bio} - [Fe^{2+}]_{mob}). \qquad (5)$$

$Fe(OH)_3$: $\dfrac{\partial}{\partial t}(\theta_{bio}[Fe(OH)_3]_{bio}) = -\dfrac{1}{U_{Fe(OH)_3}}\left[\dfrac{\partial\theta_{bio}X3}{\partial t}\right]_{grow}$

$$-\dfrac{T_{mb}(1-n)}{a}$$

$$\cdot \dfrac{\theta_{bio}\theta_{mat}\sqrt{D_M}}{\theta_{bio}+\theta_{mat}}([Fe(OH)_3]_{bio}$$

$$- [Fe(OH)_3]_{mat}). \qquad (6)$$

Matrix phase

$Fe(OH)_3$: $\dfrac{\partial}{\partial t}(\theta_{mat}[Fe(OH)_3]_{mat})$

$$= \dfrac{T_{mb}(1-n)}{a}\cdot\dfrac{\theta_{bio}\theta_{mat}\sqrt{D_M}}{\theta_{bio}+\theta_{mat}}([Fe(OH)_3]_{bio} - [Fe(OH)_3]_{mat}). \qquad (7)$$

Bacteria

X3: $\left[\dfrac{\partial X3}{\partial t}\right]_{Iron} = \left[\dfrac{\partial X3}{\partial t}\right]_{grow} + \left[\dfrac{\partial X3}{\partial t}\right]_{decay} \qquad (8)$

$\left[\dfrac{\partial X3}{\partial t}\right]_{grow} = v_{max}^{Fe(OH)_3}\cdot\dfrac{IC_{NO_3^-}}{IC_{NO_3^-}+[NO_3^-]_{bio}}\cdot\dfrac{[CH_2O]_{bio}}{K_{CH_2O}+[CH_2O]_{bio}}$

$$\cdot\dfrac{[Fe(OH)_3]_{bio}}{K_{Fe(OH)_3}+[Fe(OH)_3]_{bio}}\cdot X3. \qquad (9)$$

$\left[\dfrac{\partial X3}{\partial t}\right]_{decay} = -v_{X3dec}\cdot X3, \qquad (10)$

where θ_W, θ_{bio}, θ_{mat} are the specific volumes of mobile phases, biophase, and matrix phase. T_b, T_{mb} are the concentration change coefficients between biomobile phase or biomatrix phase, n is the porosity, a is the diameter of soil particles, $Y_{CH_2O}^{Fe(OH)}$ is the yield coefficient, $P_{Fe^{2+}}$, $U_{Fe(OH)_3}$ are the production or growth coefficients that can be stoichiometrically related to the yield coefficient. $S3_{Fe}$ is the pore water and solid phase exchange reaction term of Fe, $IC_{NO_3^-}$ is the inhibition concentration for CH_2O, K_{CH_2O}, $K_{Fe(OH)_3}$ is half saturation concentration for CH_2O, $v_{Fe(OH)_3}$, v_{X3dec} is the maximum growth or constant decay rate of bacteria X3.

3. Seawater Infiltration Experiment by Alluvial Soil with Organic Carbons

A square water tank $190\,\mathrm{cm} \times 190\,\mathrm{cm}$ with a perforated bottom was prepared as follows: coarse sand was laid onto a neutral media base in the tank to an even depth of $5\,\mathrm{cm}$ and this was covered with a layer of paddy soil, containing oxides of Fe and Mn $30\,\mathrm{cm}$ deep. The water level was maintained to a depth of $12\,\mathrm{cm}$ above the paddy soil surface. The chemical components of soil are summarized as follows (weight content (%)): Fe_2O_3; 5.34%, MnO_2; 0.05%, Al_2O_3; 8.09%, SiO_2; 77.58%, Organic matter; 6.08%, C; 1.96%, H; 0.76%, N; 0.15%. Cation exchange capacity of the paddy soil is 5.724 $(\mathrm{mmol}/100\,\mathrm{g})$. The calculation condition are as follows; calculation depth: $30\,\mathrm{cm}$, calculation period: $60\,\mathrm{days}$, grid mesh size: $0.5\,\mathrm{cm}$, time increment: $30\,\mathrm{s}$, cross sectional velocity: $1.385 \times 10^{-5}\,\mathrm{cm/s}$, $\theta_w = 48\%$, $\theta_{bio} = 2\%$, $\theta_{mat} = 50\%$, $IC_{NO_3^-} = 1.0 \times 10^{-2}\,\mathrm{mmol/l}$.

4. Application of the Numerical Simulation Model

The results of numerical simulation for the entity species after $60\,\mathrm{days}$ at $29.5\,\mathrm{cm}$ (the grid of the boundary of alluvial soil and coarse sand of the experimental setup) are shown in the right column in Table 1. The parameters used in the simulation model are listed in Table 2, respectively. Comparing with the analyzed and the calculated value of the chemical species concentration of effluent seawater concentration, it can be noted

Table 1.   Measured and calculated concentration of seawater.

Measured species	Analyzed values (after 60 days of infiltration)		Calculated effluent seawater concentration
	Infiltrated seawater concentration	Effluent seawater concentration	
DO (meq/l)	0.200	0.016	0.121
Ca^{2+} (meq/l)	21.1	20.3	20.3
Mg^{2+} (meq/l)	98.7	107.0	107.0
Na^+ (meq/l)	487.2	482.8	482.8
K^+ (meq/l)	12.3	11.6	11.6
S-Mn (meq/l)	0.000	0.011	—
T-Mn (meq/l)	0.000	0.013	0.196
S-Fe (meq/l)	0.005	0.027	—
T-Fe (meq/l)	0.426	0.423	0.651
Cl^- (meq/l)	552.9	507.7	507.7
NO_3-N (meq/l)	0.004	0.001	0.001
S-TOC (meq/l)	0.056	0.333	0.008

K. Akagi et al.

Table 2. Parameters used for the simulation model.

Biochemical parameters		
Selectivity coefficient	$K_{Na/Mg} K_{Ca/Mn} K_{Ca/Fe}{}^1$	1.1454
	$K_{Ca/Na}{}^1$	13.5625 mmol/l
	$K_{Ca/K}{}^1$	34.9903 mmol/l
Exchange coefficient	T_b	$7.7460 \times 10^{-1}\,1/s^{1/2}$
	$T_{mw}{}^7$	$1.6832 \times 10^{-6}\,1/s^{1/2}$
	$T_{mb}{}^7$	$6.8594 \times 10^{-7}\,1/s^{1/2}$
Monod constant	$K_{CH_2O}{}^7$	0.10 mmol/l
	$K_{O_2}, K_{MnO_2} K_{Fe(OH)_3}$	1.0×10^{-3} mmol/l
Aerobic bacteria X1	Yielding coefficient $Y_{CH_2O}^{O_2}{}^7$	0.1 mol cell-C/mol OC
	Maximum growth rate $v_{max}^{O_2}{}^7$	5.0 day^{-1}
	Constant decay rate $v_{X1dec}{}^7$	0.75 day^{-1}
Anaerobic bacteria X1	Yielding coefficient $Y_{CH_2O}^{NO_3^-}{}^7$	0.081 mol cell-C/mol OC
	Maximum growth rate $v_{max}^{NO_3^-}{}^7$	4.05 day^{-1}
	Constant decay rate $v_{X1dec}{}^7$	0.75 day^{-1}
Manganese bacteria X2	Yielding coefficient $Y_{CH_2O}^{MnO_2}{}^7$	0.015 mol cell-C/mol OC
	Maximum growth rate $v_{max}^{MnO_2}{}^7$	0.5 day^{-1}
	Constant decay rate $v_{X2dec}{}^7$	0.075 day^{-1}
Iron bacteria X3	Yielding coefficient $Y_{CH_2O}^{Fe(OH)_3}{}^7$	0.010 mol cell-C/mol OC
	Maximum growth rate $v_{max}^{Fe(OH)_3}{}^7$	0.5 day^{-1}
	Constant decay rate $v_{X3dec}{}^7$	0.075 day^{-1}

that DO and NO_3^- concentration decreased, while, the concentration of Mn^{2+} and Fe^{2+} increased. The concentration of the other cation show about the same. The TOC concentration in the effluent seawater increased significantly larger than the infiltrated seawater. This could be assumed by the dissolution of the organic matter contained in the alluvial soil. As compared with the analyzed and calculated concentration of effluent seawater, the calculated concentration of Ca^{2+}, Mg^{2+}, Na^+, and K^+ are nearly in accordance with the analyzed concentration except Mn^{2+} and Fe^{2+} concentrations. The calculated concentration of DO shows larger than that of the analyzed concentration. This phenomenon can be assumed that the available organic carbon for bacteria X1 is limited, therefore, bacteria X1 does not grow. In the next place, the calculated concentration of Fe^{2+} and Mn^{2+} shows lager than these of the analyzed concentration. It was inferred that Fe^{2+} reacted with sulfide ion as was seen for the precipitation of iron sulfide in the experimental apparatus. Finally, the calculated concentration of organic carbon is smaller than that of the analyzed concentration, from consideration of only dissolution the available organic carbon in numerical simulation.

5. Calculated Results and Conclusion

Figure 2 shows the time series of $Fe(OH)_3$ concentration in matrix phase at the depth of 5 cm. Figures 3 and 4 show the time series of $Fe(OH)_3$ concentration or Fe^{2+} concentration in biophase, Fig. 5 shows the time series of Fe^{2+} concentration in mobile phase. Figure 6 shows the time series of CH_2O and bacteria X3 in biophase. These figures show that the concentration of iron hydroxide in matrix phase decreases. While, the concentration of iron hydroxide in biophase increases. Through the bacteria-mediated reduction divalent iron in mobile phase changes corresponding to the concentration in biophase. Since bacteria cannot grow without sufficient organic carbon, the concentration of divalent iron does not increase in biophase. Thus, the concentration of divalent iron in mobile phase is controlled by mass-transfer reaction.

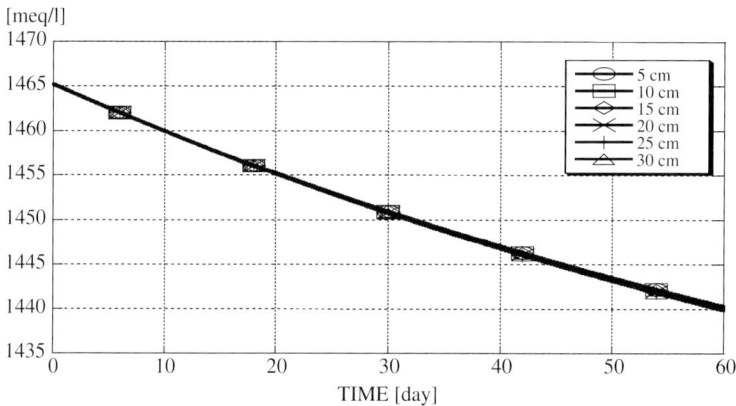

Fig. 2.   Time series of iron hydroxide in matrix phase.

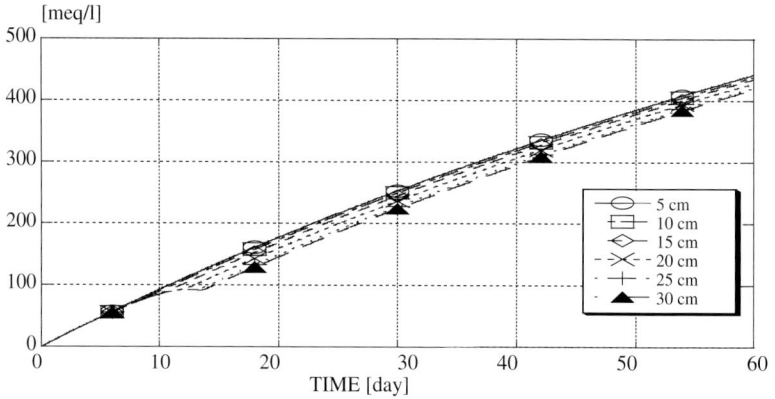

Fig. 3.   Time series of iron hydroxide in biophase.

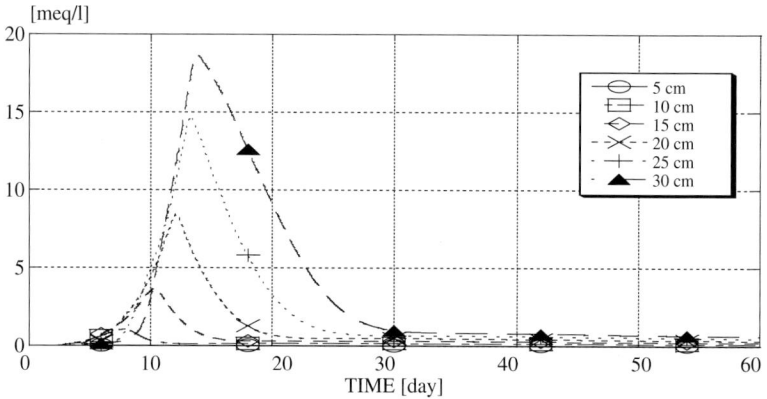

Fig. 4.   Time series of divalent iron in biophase.

The concentration of divalent iron in both biophase and mobile phase reaches a peak at about the 13th day. At this time, the concentration of bacteria X3 also reaches a peak, organic carbon is almost consumed out. In Fig. 3, it seems that all the concentrations of iron hydroxides in biophase along any depth approaches with each other for the first 8 days, while for the subsequent time the slightly deviation took place in the lower layer. This is because that solute organic carbon in the upper layer is transported downward and stored into the lower layer. As a result, organic carbon and

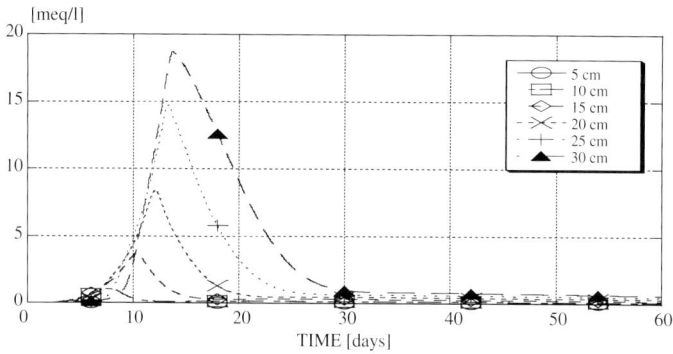

Fig. 5. Time series of divalent iron in mobile phase.

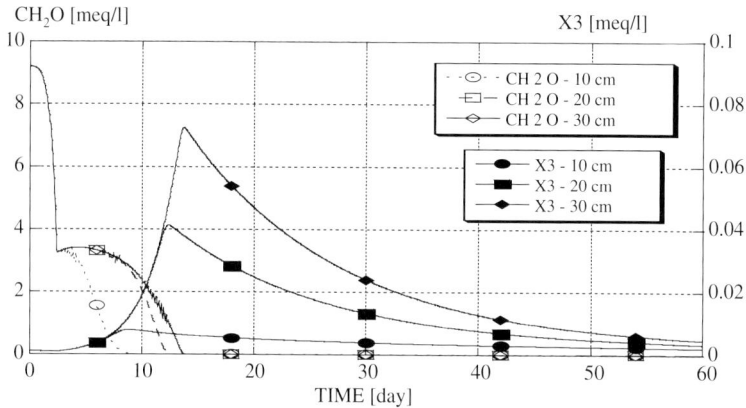

Fig. 6. Time series CH_2O and X3 in biophase.

bacteria X3 increase in biophase yield more iron hydroxide into divalent iron in the lower layer.

The model describing the groundwater quality in a coastal aquifer was studied through the comparisons with the experiment by the one-dimensional column of alluvial soil. The model calculates the mass transport and cation exchange and bacteria-mediated biochemical reducing processes, specifically the main focus was put on the inclusion of reduced divalent manganese and iron ions into the cation exchange reactions. The approach was considered useful to predict the reduced level of the seawater in a coastal aquifer. The applicability of the present approach needs to be calibrated in the field.

References

1. G. Guerra, K. Jinno and Y. Hiroshiro, Behavior of chemical species under redox environment using multi-component solute transport model, *Annual Journal of hydraulic Engineering* **47** (2003) 319–324.

2. W. Kinzelbach, W. Schäfer and J. Herzer, Numerical modeling of natural and enhanced denitrification process in aquifer, *Water Resources Research* **27**, 6 (1991) 1149–1159.

3. H. J. Lensing, M. Vogt and B. Herrling, Modelling of biologically mediated redox processes in the subsurface, *Journal of Hydrology* **159** (1994) 125–143.

4. D. Schäfer, W. Schäfer and W. Kinzelbach, Simulation of reactive process related to biodegradation in aquifers 1. Structure of the three-dimensional reactive transport model, *Journal of Contaminant Hydrology* **31**, 1 (1998) 167–186.

5. D. Schäfer, W. Schäfer and W. Kinzelbach, Simulation of reactive process related to biodegradation in aquifers 2. Model application to column study on organic carbon degradation, *Journal of Contaminant Hydrology* **31**, 1 (1998) 187–209.

6. K. Momii, Y. Hiroshiro, K. Jinno and R. Berndtsson, Reactive solute transport with a variable selectivity coefficient in an undisturbed soil column, *Soil Science Society of America Journal* **61**, 6 (1997) 1539–1546.

7. Y. Hioroshiro, K. Jinno, T. Yokoyama and M. Kubota, Multicomponent solute transport with cation exchange in a redox subsurface environment, *Calibration and Reliability in Groundwater Modelling, Proceeding of the ModelCARE99*, pp. 474–480, 1999.

8. M. Snyder, M. Taillefert and C. Ruppel, Redox zonation at the saline-influenced boundaries of a permeable surficial aquifer, *Journal of Hydrology* **296** (2004) 164–178.

CALCIUM (Ca²⁺) DETERIORATION UNDER THE SUBSURFACE RUNOFF IN A FORESTED HEADWATER CATCHMENT

KASDI SUBAGYONO[*,‡], TADASHI TANAKA[†] and MAKI TSUJIMURA[†]

Center for Soil and Agroclimate Research and Development
Jl. Ir. H. Juanda, 98 Bogor 16123, Indonesia
†*Graduate School of Life and Environmental Sciences, University of Tsukuba*
Ibaraki 305-8572, Japan
‡*kasdi_s@yahoo.com*

The present study deals with the nature of Ca^{2+} under both weathering and transport processes in Kawakami forested headwater catchment, central Japan. The study showed that Ca^{2+} deterioration is prominently occurred from which the subsurface runoff governing transport of Ca^{2+} is dominance, the process by which the Ca^{2+} is leached out. Shallow riparian groundwater (near surface riparian) is the most dominant source area which releases much concentration of Ca^{2+}. The rate of Ca^{2+} leach out from this source area has been estimated about 5.5 mg Ca^{2+} ha/day (or 0.002 kg Ca^{2+}/ha/year). Ca^{2+} deterioration was much obvious when the input through weathering is lower to that of the output through the transport process.

1. Introduction

Insufficient rate of weathering with respect to the rate of transport under the subsurface runoff may create Ca^{2+} to deteriorate in a forested headwater catchment. Yet, the study on Ca^{2+} balance between one created from weathering and its loss through transport process is somewhat rare.

Ca^{2+} may deteriorate through several processes in the soil such as leaching[1-3] and vegetation uptake.[3] Mulholland et al.[4] reported that exports of most solutes (includes Ca^{2+}) from the watershed exceeded precipitation inputs leading to a negative balance of Ca^{2+} in the soil. Ca^{2+} is dominant chemical mixture in the stream channel and it was the product of weathering in the catchment studied.[5]

The present study is focused to (a) elucidate the relationship between hydrologic flowpath and Ca^{2+} pathway, (b) assess Ca^{2+} transport with respect to its weathering, and (c) assess the Ca^{2+} depletion rate within headwater catchment.

2. Site Description

The study area is Kawakami headwater catchment, 14 ha area situated at the center-west of Nagano prefecture, Central Japan (35°54.9′N, 138°30.2′E) (Fig. 1). About 5.2 ha from the total area were used for the experiment. The altitude of the catchment ranges from 1,500 to 1,680 m asl with slopes range from 20% to > 60%. This area is underlied by Late Neogene of the Meshimoriyama volcanic rocks.[6] The upper soil mantel primarily consists of Inceptisols. The A-horizon has a rapid hydraulic conductivity ($Ks = 21.6 - 93.6$ cm/h), while B-horizon has very slow hydraulic conductivity ($Ks = 0.007$–0.9 cm/h). Ca^{2+} concentration in shallow riparian groundwater, deep riparian groundwater and hillslope soil water during baseflow period was 4.81, 7.88, and 0.74 mg/l, respectively. Mean annual precipitation is 1,500–1,600 mm, producing 800–900 mm of runoff.

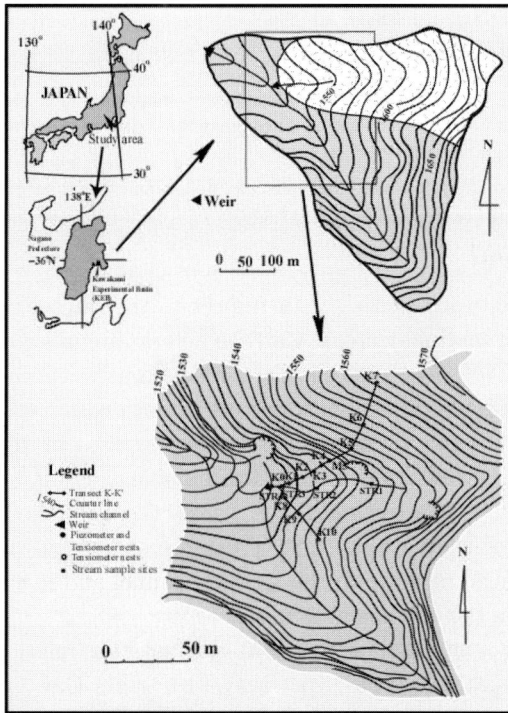

Fig. 1. Map of Kawakami headwater catchment showing the transect **K-K′**.

3. Methods

The study has been monthly monitored for 13 months from August 2000 to August 2001 and during the storm event on August 21–22, 2001. A nested transect was used to understand the mobility of Ca^{2+} across hill slope and riparian zone. Discharge was continuously recorded at 30°V-notch gauging weir and rainfall was measured using tipping bucket (recording) rain gauge.

Groundwater, soil water, and stream water samples were collected prior for Ca^{2+} measurement. Groundwater samples were taken from the piezometer, whereas soil water samples were collected from suction samplers installed at the same site with piezometer and tensiometer nests. Stream water was taken at four different sites (Fig. 1).

The water samples were filtered through $0.22\,\mu m$ Millipore membrane filters to separate suspended matter and the filtered solutions were analyzed for Ca^{2+} and SiO_2. Ca^{2+} and SiO_2 concentrations were measured using Inductive Couple Argon Atomic Emmision Spectro-photometer at the Chemical Analysis Center of University of Tsukuba.

Ca^{2+} flux was quantified to assess Ca^{2+} transport using the convection-dispersion model as follows:[7]

$$J_l = J_w C_l - D_{lh}\partial C_l/\partial z - D_l^s \partial C_l/\partial z, \tag{1}$$

where J_l is the total flux of dissolved Ca^{2+}; J_w is the water flux; C_l is dissolved Ca^{2+} concentration; D_{lh} is the hydrometric dispersion coefficient; z corresponds to the distance where the Ca^{2+} is transported; and D_l^s is the soil liquid diffusion coefficient.

$$J_w = -Ks(\partial H)/(\partial z), \tag{2}$$

$$D_{lh} = \lambda V, \tag{3}$$

$$V = J_w/\theta, \tag{4}$$

$$\lambda = 0.0169 Ls^{1.53}, \tag{5}$$

$$D_l^s = \xi(\theta) \cdot D_l^w, \tag{6}$$

$$\xi(\theta) = \theta^{10/3}/\phi^2, \tag{7}$$

where J_l is the total flux of dissolved solute (mg/cm/se); J_w is the water flux (cm/s); C_l is dissolved solute concentration (mg/l); D_{lh} is the hydrometric dispersion coefficient (cm²/s); z corresponds to the distance where the solute is transported (cm); D_l^s is the soil liquid diffusion coefficient (cm²/s), D_l^w is the diffusion coefficient of the solute in water (cm²/s), Ks is saturated hydraulic conductivity (cm/s), $(\partial H)/(\partial z)$ is hydraulic gradient (cm/cm), λ is dispersivity (cm) which was adopted from Ref. 8, V is pore water velocity (cm/s), θ is volumetric water content (cm^3/cm^3), Ls is the apparent length scale which is correspond to the distance (m), $\xi(\theta)$ is the tortuosity, and ϕ is the porosity.

Assessment of Ca^{2+} input from weathering has been made using the mineralogical data from the X-ray deffragtograms for primary and secondary minerals in the soil. The trend of its concentration in both soil and water (groundwater, soil water) was used to synthesize and analyze its balance.

4. Results and Discussion

4.1. *Weathering and concentration of Ca^{2+}*

As high as 6.07% of CaO has been measured in the top 20 cm of soil in the riparian zone in which it was much higher than that in 80 cm depth (1.5%). In contrast, during base flow Ca^{2+} concentration in the shallow (0.2–0.6 m) groundwater was as high as 4.81 mg/l which significantly lower as that in the deep (1–2 m) groundwater (7.99 mg/l), see Table 1. Stream Ca^{2+} concentration decreased during the rising limb to reach the lowest concentration at near peak runoff (Fig. 2).

The XRD analysis has shown that quartz (SiO$_2$) is the most abundant mineral in the soil, followed by feldspar which is mainly anorthite (CaAl$_2$Si$_3$O$_8$), slightly rare gibbsite (Al(OH)$_3$), and very rare albite (NaAlSi$_3$O$_8$) (Fig. 3). In general, these minerals were distributed homogeneously amongst the soil horizons, except for the upper horizon of the riparian zone.

In the groundwater and soil water, SiO$_2$ and Ca^{2+} concentrations were dominance suggesting that these solutes are originated from these minerals through the weathering. However, inconsistency was identified in which it was a trend in increasing Ca^{2+} concentration in the groundwater with depth, but not the case in the soil profile. This provides insight into the process by which Ca^{2+} especially in the upper soil solution was leached out through subsurface runoff. This phenomenon also suggests that the rate of weathering is inferior to the rate of transport, which leads Ca^{2+} concentration to deteriorate.

4.2. *Ca^{2+} transport*

It has been reported that the near surface riparian/NSR (45%), deep riparian groundwater/DRG (20%) and hillslope soil water/HSW (35%) are the major sources area contributing to runoff generation in this catchment.[5]

Table 1. Ca^{2+} concentration in source area of runoff and stream during storm event on August 21–22, 2001.

Sources of runoff	Ca^{2+} (mg/l) Mean ± SD (n)	Stream Ca^{2+} (mg/l)
Baseflow		
1. Near surface riparian	4.81 ± 1.74 (8)a	8.37
2. Deep riparian groundwater	7.99 ± 2.81 (12)b	
3. Hillslope soil water	0.74 ± 0.67 (20)c	
2 h after started		
1. Near surface riparian	4.65 ± 1.72 (8)a	7.18
2. Deep riparian groundwater	7.79 ± 2.67 (12)b	
3. Hillslope soil water	0.78 ± 0.71 (16)c	
1.5 h after peak		
1. Near surface riparian	4.86 ± 1.70 (8)a	4.79
2. Deep riparian groundwater	8.13 ± 2.81 (12)b	
3. Hillslope soil water	0.70 ± 0.26 (16)c	
Storm end		
1. Near surface riparian	4.68 ± 1.68 (8)a	4.56
2. Deep riparian groundwater	8.30 ± 2.74 (12)b	
3. Hillslope soil water	0.66 ± 0.61 (23)c	
Post storm		
1. Near surface riparian	4.75 ± 1.72 (8)a	7.37
2. Deep riparian groundwater	8.50 ± 3.54 (12)b	
3. Hillslope soil water	0.73 ± 20.62 (23)c	

Mean values in the same column with the same letter are not significantly different based on one-way ANOVA ($P < 0.0001$) and multiple comparison tests ($\alpha = 0.05$, Tukey). SD: standard deviation; n: number of data.

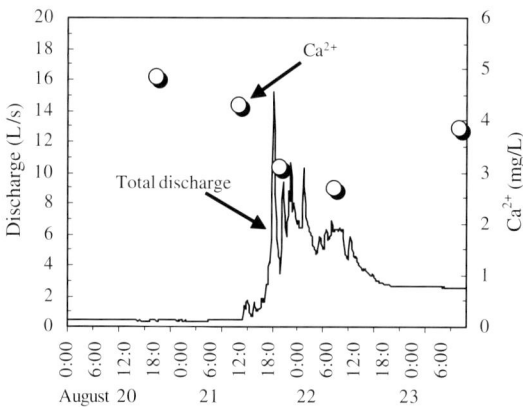

Fig. 2. Ca^{2+} concentration across the hydrograph during storm event on August 21–22, 2001. NSR: near surface riparian; DRG: deep riparian; groundwater and HSW: hillslope soil water.

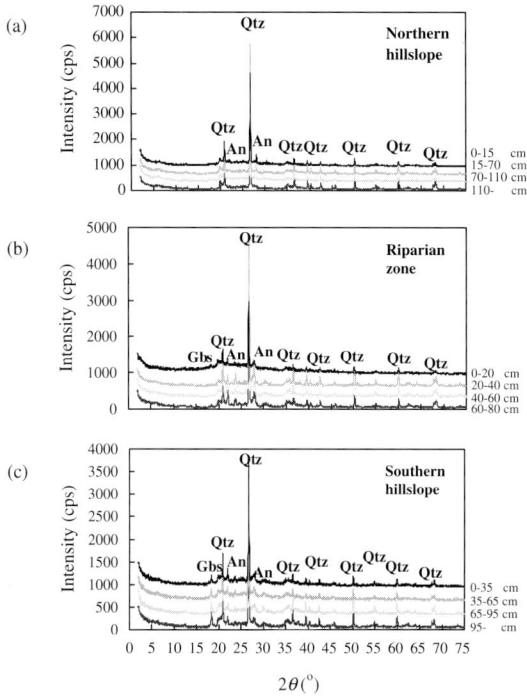

Fig. 3. X-ray diffractograms for primary minerals in (a) northern hillslope, (b) riparian zone, and (c) southern hillslope. Qtz: quartz; An: anorthite; Gbs: gibbsite.

There was positive correlation with good agreement ($R^2 = 0.85$) between the relative contribution of NSR to runoff generation and the change of the stream Ca^{2+} concentration during storm event on August 21–22, 2001 (Fig. 4(a)). Stream Ca^{2+} concentration increased with the increase NSR contribution to runoff generation. The mixing diagram (Fig. 4(b)) shows that the stream chemistry was bounded by those three end-members. Stream Ca^{2+} concentration was much similar to that of the NSR water at base flow condition then changed toward HSW when the storm started and progressively changed toward peak storm until the end of storm and returning to the value for NSR at the post storm. This suggests that Ca^{2+} in the NSR transported to the stream channel by the lateral flow. During this large storm, about 5.5 mg Ca^{2+}/ha/day (or 0.002 kg Ca^{2+}/ha/year) has been leached out from the NSR.

In contrast, the higher concentration of Ca^{2+} in the top 20 cm of soil, compared with the deeper layer, was not in positive correlation with that

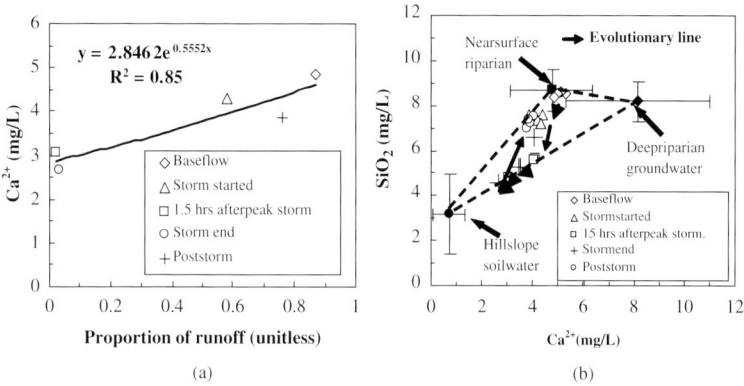

Fig. 4. Relationship between contribution of NSR to runoff and change of stream Ca^{2+} concentration (a) and mixing diagram of Ca^{2+} and SiO_2 during August 21–22, 2001 storm.

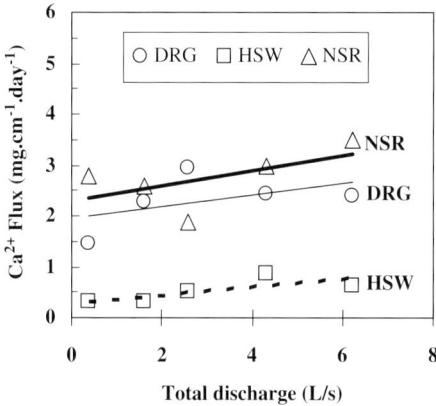

Fig. 5. Relationship between discharge and Ca^{2+} flux during storm event on August 21–22, 2001. NSR: near surface riparian; DRG: deep riparian groundwater; HSW: hillslope soil water.

in the NSR suggesting that the rate of transport was superior to that of dilution of the weathered Ca^{2+}.

Those phenomena explain well that Ca^{2+} has progressively deteriorated during the hydrologic event. As discharge increased Ca^{2+} flux increased linearly (Fig. 5). The Ca^{2+} flux in the NSR was higher than that in the DRG as well as in the HSW. In the riparian zone, Ca^{2+} concentration increased away from the stream and decreased at the border between the

68   *K. Subagyono et al.*

hillslope and the riparian zone. It was progressively increased with depth. This evidence showed that source of Ca^{2+} in the stream channel has been dominated by that discharging from the NSR. The processes suggest that the dynamic of Ca^{2+} has been much determined by the flow process rather than that defined by the dilution process.

5. Conclusions

The hydrometric, Ca^{2+} concentration and mineralogical data of the catchment studied suggest that (a) Ca^{2+} deterioration is prominently occurred from which the subsurface runoff governing transport of Ca^{2+} is dominance, the process by which the Ca^{2+} is leached out, (b) shallow riparian groundwater (near surface riparian) is the most dominant source area, which releases much concentration of Ca^{2+}, and (c) Ca^{2+} deterioration will be much obvious when the input through weathering is lower to that of the output through the transport process.

This study provides insight into the role of the major subsurface flow on the potential leaching of conservative solutes in a headwater catchment. The nature of weathering and solutes dilution determines their mixture generating the stream chemistry. It is urgently needed to expand the study into the topic of biogeochemical behavior rather than that of hydrochemistry.

Acknowledgments

he study has been funded by the Ministry of Education, Science, Sport and Culture of Japan (MONBUSHO). Thanks are extended to Prof. Dr. Norio Tase and Dr Michiaki Sugita for valuable suggestions and Dr. Tamao Hatta for his assistance on XRD analysis.

References

1. D. W. Johnson and D. E. Todd, Nutrient cycling in forests of Walker Branch Watershed, Tennessee: role of uptake and leaching in causing soil changes, *J. Environ. Qual.* **19** (1990) 97–104.
2. K. Ohrui and M. J. Mitchell, Hydrological flow paths controlling stream chemistry in Japanese forested watersheds, *Hydrol. Process* **13** (1999) 877–888.
3. T. G. Huntington, R. P. Hooper, C. E. Johnson, B. T. Aulenbach, R. Cappellato and A. E. Blum, Calcium depletion in a southeastern United States forest ecosystem, *Soil Sci. Soc. Am. J.* **64** (2000) 1845–1858.
4. P. J. Mulholland, G. V. Wilson and P. M. Jardine, Hydrogeochemical response of a forested watershed to streams: effect of preferential flow along shallow and deep pathways, *Water Resour. Res.* **12** (1990) 3021–3036.

5. K. Subagyono, T. Tanaka, Y. Hamada and M. Tsujimura, Defining hydro-chemical evolution of stream flow through flowpath dynamics in Kawakami headwater catchment, Central Japan, *Hydrol. Process* **19** (2005) 1939–1965.
6. S. Kawachi, Geology of the Yatsugatake District, *Regional Geological Report*. Geological Survey of Japan, 89–91 (in Japanese with English abstract), 1977.
7. W. A. Juri, W. R. Gardner and W. H. Gardner, *Soil Physics*, 5th ed. (John Willey & Sons, Toronto, 1991) 328p.
8. S. P. Neuman, Universal scaling of hydraulic conductivities and dispersivities in geologic media, *Water Resour. Res.* **26** (1990) 1749–1758.

GEOCHEMISTRY OF GROUNDWATER AND THEIR RELATIONSHIP WITH GEOLOGICAL FORMATIONS

V. K. SAXENA*, N. C. MONDAL and V. S. SINGH

National Geophysical Research Institute
Hyderabad 500 007, India
*vks_9020010@yahoo.co.in

Studied area is a tribal area of Warangal District, Andhra Pradesh (A.P.), India. This is about $1,500\,km^2$ and has different types of rock formations such as granite gneisses, sandstones, pakhals, and alluvium, etc. This area was facing groundwater problems since two decades. A large number of shallow bore wells were drilled. These bore wells are discharging water from 10,000 to 25,000 l/h. A large number of groundwater samples were collected and quantitatively analyzed. The results indicate that: (1) These groundwaters are classified as $Na–Ca–Cl–HCO_3$ and mixed water types. (2) Fluoride is more in groundwater of granite gneisses areas. (3) Concentrations of aqueous ionic species have changed with different rock formations. (4) Groundwaters of granite gneisses areas are more in Total Dissolved Solids (TDS). (5) TDS of groundwater and depths of bore wells have shown a good correlation.

1. Introduction

Groundwater is valuable only when its quality is suitable for the purpose for which it is being explored. Suitability of groundwater for particular purpose depends upon the standards of acceptable quality for that use. The acceptability of groundwater for a particular usage is essentially dictated by the standards of acceptable/permissible limits.[1–4] Warangal, which is located in the southeast parts of India, has been facing groundwater problems in term of quantity and quality. This is part of forest and developed for the habitation of tribals. Their main source of living is based on agriculture. Because of the lack of groundwater or surface water, these people are facing various types of problems for their survival. State Government had drilled a large number of bore wells (>40 bore wells, 10–70.5 m depths). It is interesting to note that this region is having different types of rock formations, which is Recent to Archaean.[5] For the study of chemical quality of these groundwaters in this area, the water samples were collected from the locations of these bore wells, which are shown in Fig. 1. This paper evaluates (1) quality of groundwater in relation to potable purpose, (2) changes

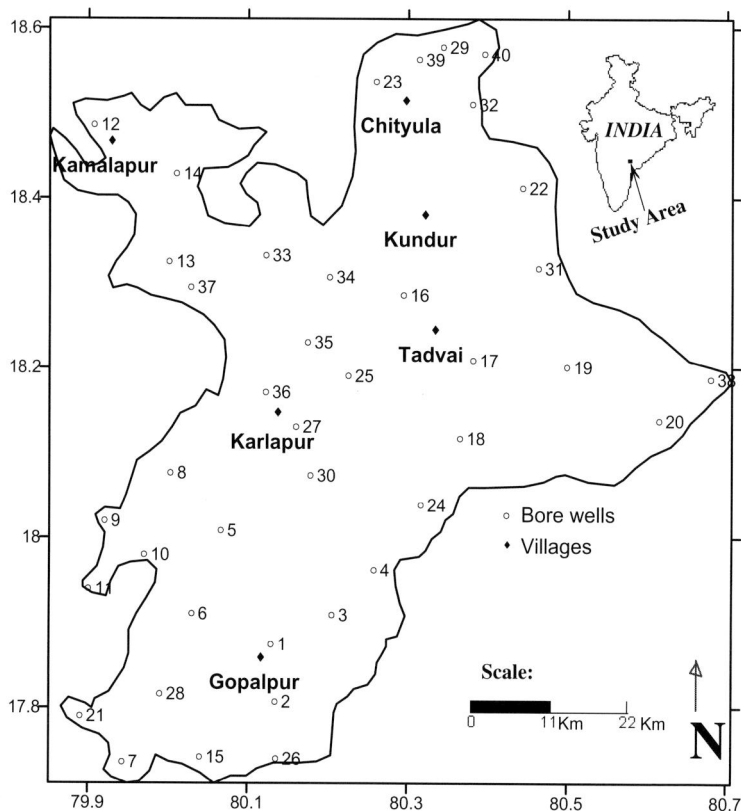

Fig. 1.   Location of bore wells, Warangal, A.P., India.

of hydrochemical constituents with rock formations, and (3) variations of aqueous ionic species with depth of bore wells.

2. Geology and Hydrogeological Setting

The geological features of this area are shown in Fig. 2. This area is comprised of different types of geological formations such as (1) Archaeans, (2) Puranas, (3) Gondwanas, and (4) Recent, etc. The Archaeans are basically granites, granite gneisses, schists and dolerites, etc. Puranas are principally of sedimentary and consist of pakhals and sulavaries. Pakhals are equivalent to Cuddapah consisting of slates, phyllites, dolomites, quartzites, and shales, etc. Gondwanas are mainly of Lower Gondwana and have

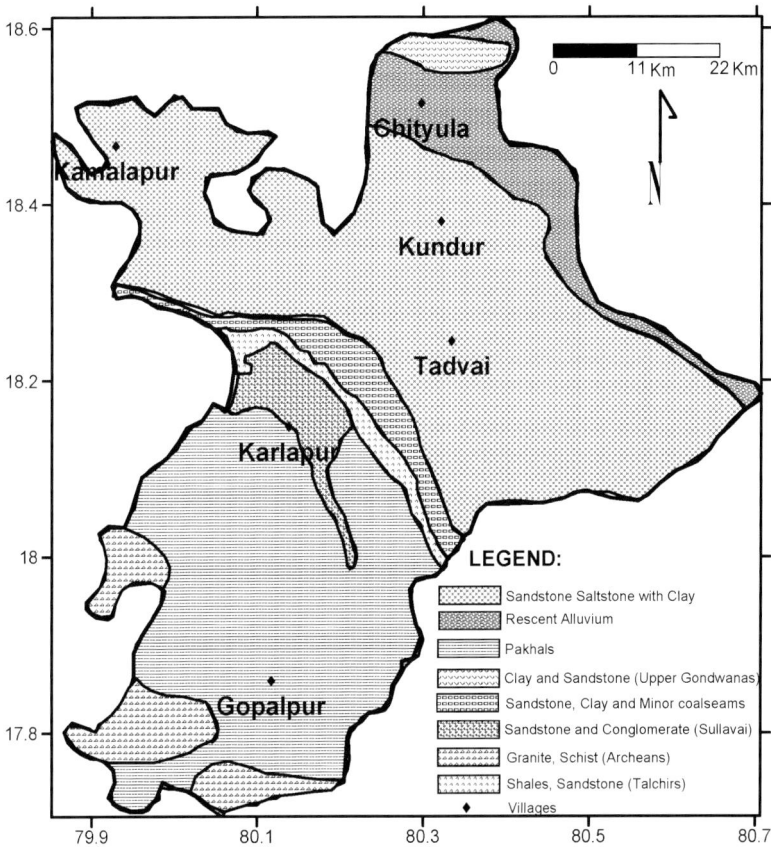

Fig. 2.   Geological map, Warangal, A.P., India.

Talchirs followed by barakars, sandstones, and kamthis. Gondwanas occupied about 50% of the study area.[5]

This area is having valley, hills, and plain lands. Elevation varies from 70 to 160 m (amsl.). Climate of this area is tropical semi-arid. Average surface temperature is 28°C. Annual rainfall is 900–1,300 mm.[5] About 80% rain occurs during monsoon season, which is usually active from July to October.

Godavari river flows along the eastern boundary of this area, in ES-direction and at a distance of 180 km. Groundwater in these formations generally occurs under phreatic condition particularly in shallow weathered zones. Some times, it occurs under semi-confined and confined conditions at deeper fracture zones.[5] Average depths of weathered zone vary from 8 to 18 m. Groundwater table lies in between 5 and 15 m and mostly depend upon rock formations. Main soils are silts, black cotton soils, clay loams and sandy soils. Main crops are paddy, jawar, maize, cotton, chilies, tobacco, and pulses, etc.

3. Water Sample Collection and Analytical Techniques

Water samples were collected in 1-liter polythene bottles and the standard instrumental/chemical techniques were used for water samples analysis.[6–8]

4. Results and Discussions

Chemical analyses of these water samples are shown that the groundwaters are nearly neutral to mildly alkaline in pH (7.2–8.2). Electrical conductivity varies from 328 to 1,610 μS/cm, and shows a big variation. In 25% samples total dissolved solids (TDS) are not within the permissible limit of potable/domestic water.[2] The water emerges from bore well Nos. 9, 10, 11, 12, 15, 21, 26, and 28 have shown enrichment of Na, Ca, Cl, HCO_3, and F in water which is more than the permissible limit of drinking water.[2] Maximum electrical conductivity (1,610 μS/cm) was found in Govindraopet (bore well No. 10), which is located in rock area and minimum in Khanepur (328 μS/cm) in the bore well No. 32, where geological formation is recent. Similarly, high Na and Ca were found in bore wells Nos. 21 and 9 both are in granites and lowest in bore wells Nos. 40 and 32 are in recent. Maximum Mg (45 mg/l) was determined in Gudur (bore well No. 26 in Upper Gondwanas) and minimum in Khanepur, alluvium (5 mg/l). Thus, it is found, groundwaters with low TDS are associated with alluvium. Govindraopet groundwater is also enriched in HCO_3 and chloride, which shows the higher dissociation rate in hard rocks. The variability of aqueous ionic species of groundwater with different types of rock formation, may indicate, the possibility of changing the hydrochemistry with geological formations. Based on chemical analysis data, these groundwaters have been classified as: Na–Ca–Cl–HCO_3 and mixed types. Chemistry of these bore wells and their locations, indicated that groundwater emerges in sandstone and alluvium are mostly have low salinity, whereas shales and clay have medium to high; and granite gneisses with high salinity. However, the concentrations of cations (Ca, Mg, Na, and K) and anions (Cl, HCO_3 and SO_4), which gives preliminary information about the water quality and

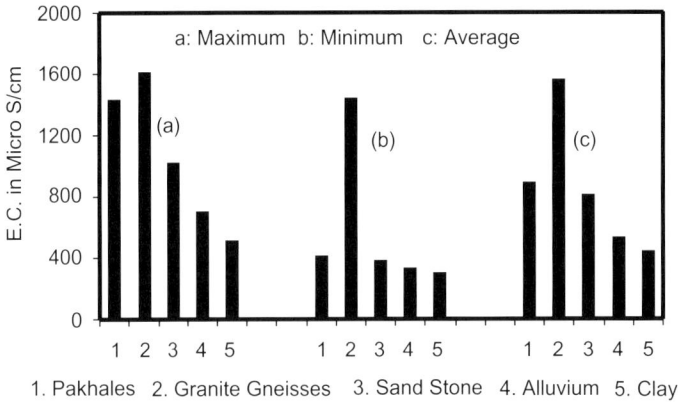

Fig. 3.   EC of bore well waters in different geological formations.

its suitability are more in groundwater of hard rock areas as compared to sedimentary or recent. The more emphasis has been given for the detailed chemical study of these bore wells water. Based on chemical composition of bore well water and associated geological formations (Fig. 2), these groundwaters have been studied. It is observed; more or less some relationship has been observed in between geological formations and water types. The EC is more in granite gneisses compared to recent and clay. The results are presented in Fig. 3. This figure showed the changes in EC with geological formations. Such changes are also found to be correlated in term of maximum, minimum and average values of EC. For examples, higher EC levels are shown in granite gneisses against low in clay. The changes in bore well water chemistry with depths have been studied. To make this study more simple the bore wells are divided into two categories (1) shallow (<50 m depth) and (2) deep (>50 m depth). The TDS of bore well waters and the depths of bore wells have been examined and found to have a positive correlation. In addition to this, those bore wells are located in granite gneisses may have deeper depths and less discharge. The data showed that 86% bore wells in hard rock areas may have deeper depths (>50 m). A fairly good discharge (>20,000 l/h) has been shown in sedimentary formations.

Fluoride concentration varies from 1 to 2.5 mg/l, 25% samples of the total have shown fluoride, which is more than permissible limit of drinking water.[2] It is also indicated that the high F contents are observed in Govindropet (sample No. 10) and Bhimla near sample No. 15. These locations are within the granite gneisses. However, the high fluorite (0.4–0.9%)

is reported in the granites gneisses in some areas of India.[9–11] The fluoride contents have been plotted with calcium and indicated the positive correlation.

Acknowledgments

Authors wish to thank to Dr. V. P. Dimri, Director, NGRI, Hyderabad, India, for his permission to publish this paper.

References

1. American Water Works Association, McGraw Hills, New York, 654 (1971).
2. World Health Organization, Guideline to drinking water quality, World Health Organization, Geneva, 186 (1983).
3. V. K. Saxena, in *Water Ecology, Pollution and Management*, eds. S. Rao and S. Pitchaiah (Chugh Publication, Allahabad, India, 1991), pp. 125–147.
4. V. V. Kumar and V. K. Saxena, *Proc Int. Symp. Applied Geochemistry* (1996) 273–279.
5. Ground Water Technical Report, A report on the status of groundwater investigations and scope for further groundwater development in the tribal subplan area of Warangal District, A.P., 30, 1987, pp. 1–120.
6. APHA, *Standard Methods for the Examination of Water and Waste*, 16th edition (APHA, Washington DC, 1985), pp. 1–50.
7. E. Browen and M. J. Skougstad, US Government Printing, Washington, 1974, pp. 1–160
8. V. K. Saxena and S. Ahmed, *Environ. Geol.* **40** (2001) 1084–1087.
9. V. K. Saxena and S. Ahmed, *Environ. Geol.* **43**, 6 (2003) 731–736.
10. N. C. Mondal, V. K. Saxena and V.S. Singh, *Curr. Sci.* **88**, 12 (2005) 1988–1994.
11. N. C. Mondal, V. K. Saxena and V. S. Singh, *Environ. Geol.* **48**, 2 (2005) 149–157.

DYNAMIC AND CHEMICAL EVOLUTION OF GROUNDWATER SYSTEM IN QUATERNARY AQUIFERS OF YANGTZE ESTUARINE REGION, CHINA

BAOPING SONG*, ZHONGYUAN CHEN[†] and ZHENG FANG[‡]

*Department of Resources and Environment, Shijiazhuang College
Shijiazhuang, China
[†]Department of Geography, East China Normal University
Shanghai, China
[‡]Shanghai Institute of Geological Survey, Shanghai, China

As a part of groundwater flow system of Yangtze Delta plain, the groundwater in Yangtze estuarine region flows much slowly due to closed circumstance. However, during the last several decades, the groundwater dynamic environment has been changed seriously, especially for the II-aquifers, because of heavy groundwater withdrawal and artificial recharge. In order to reveal the evolution pattern of groundwater system under natural condition and human activities, some geochemical methods and groundwater flow model are applied in this study, supported by a large database on groundwater chemistry, artificial discharge/recharge, and hydrogeology. The results show that three groundwater flow systems can be recognized. Under the natural condition, groundwater evolves in manner of dilution by rainwater in shallow system but in manner of mixing and cation exchange in middle system, and in manner of cation exchange in deep system.

1. Introduction

The Yangtze Delta plain, one of the most important economic zones in China, is located in the eastern coast of China. The delta is composed of Quaternary sediments with a big thickness of 100 m in the western plain and more than 400 m in the estuarine depocenter,[1] where abundant groundwater resources have provided important support for economic development. Shanghai region, occupied major area of the estuarine region, is this study area.

The previous hydrogeological survey[2] distinguished one phreatic aquifer and five confined aquifers in the estuarine region. The confined aquifers were marked, respectively, as I–V downward shown briefly in Table 1.

Most previous studies started from early 1950s and focused primarily on groundwater resources assessment, groundwater flow model limited in

Table 1.   The basic characteristics of all aquifers in the study area.

Aquifer	Time	Thickness (m)	Composition	SWG* (m^3/day)
Phreatic aquifer	Holocene	5–20	Fine sand, silt, and clayey silt	<319.5
I-aquifer	Late Pleistocene	6–15	Silty fine sand, silt	500–1000
II-aquifer	Late Pleistocene	20–40	Fine sand, gravel medium to fine sand	3000–5000
III-aquifer	Middle Pleistocene	20–30	Silty fine sand, medium to fine sand and sandy gravel	3000–5000
IV-aquifer	Early Pleistocene	30–50	Fine sand, coarse to medium sand, sandy gravel	100–3000
V-aquifer	Early Pleistocene	10–50	Coarse to medium sand, sandy gravel	100–3000

*SWG is the specific well discharge.

range of several wells or small area, and ground subsidence due to ground-water overdraft. Little has been known for regional dynamic and chemical evolution under natural condition and human activities. The present study emphasizes particularly on groundwater chemical analyses, combining with water flow modeling, to reveal the ways of hydrochemical evolution and the change of groundwater flow field affected by human activities.

2. Database and Methodology

The database used for hydrochemical and hydrogeological analyses is established on the basis of 78 Quaternary hydrological boreholes, which were drilled during 1958–1985 by the Geology and Mineral Bureau of Shanghai (GMBS). Totally, 259 water samples were collected from five confined aquifers, of which 31 samples were from I-aquifer, 70 from II-aquifer, 53 from III-aquifer, 70 from IV-aquifer, and 35 from V-aquifer. The concentration of $K^+, Na^+, Ca^{2+}, Mg^{2+}, NH_4^+, Al^{3+}, Cl^-, HCO_3^-, CO_3^{2-}$, and SO_4^{2-} in groundwater samples were analyzed by the chemical analysis laboratory of GMBS, and the values of pH and temperature were also measured.

To describe the variety of groundwater flow influenced by human activities, a quasi-three-dimension numerical model is established using MOD-FLOW software of US Geological Survey.[3] The phreatic aquifer and aquifers I–III were included in the model. So a great deal of information about hydrogeology of four aquifers are collected, together with water heads in 157 observation wells and the yields of 1,527 artificial discharge/recharge wells in 1980–1990.

The study area is divided uniformly into 12,000 spaced grids. Each grid cell represents $1,127 \times 1,228\,\mathrm{m}^2$. Lateral exchange of groundwater with the region immediately adjacent to the study area is adjusted by the general head boundary of the MODFLOW according to the water table difference between inside and outside of the boundary.

Simulating stage is from 1980 to 1989, and the water heads in 1990 are used for subsequent prognoses. By reason of records of groundwater pumping and injection go by months, the simulating stage is divided into 12 stress periods. Each stress period consists of two time steps.

3. Hydrochemical Evolution of Groundwater

Long-term groundwater regime reveals that water table in phreatic aquifer is controlled mainly by micro-geomorphology and rainfall, and rainwater is the most important recharge source. The composition of hydrogen isotopes supports this view that tritium concentration of groundwater in 1980 (21.3–35.6 tritium units (TU)) is similar to that of rainwater (30 TU).[4] Furthermore, three facts can be found from Fig. 1: (a) saltwater bodies distribute inside transgression scope of early Holocene, (b) groundwater samples with the highest TDS distribute in east coast, and (c) the groundwater ridge is approximately accordant with a high belt of geomorphology in south of estuary. These indicate that the saltwater in phreatic aquifer was primitively formed from seawater in Holocene and was diluted by rainwater subsequently.

From Fig. 1(b), the distribution of water quality in confined aquifers is characterized obviously by: (a) increase of saltwater area and gradual decrease of freshwater area from lower to upper; (b) expansion of saltwater area from west to east region of the study area, and (c) decrease landward of TDS values. Sedimentologic, paleontologic and geologic investigation demonstrated that there were 4–7 times transgression events during Quaternary in the estuarine region and the intensity of these events tended to increase from early Pleistocene to Holocene.[6,7] It is thus to believe that the temporal and spatial distribution of water quality was generally controlled by the Quaternary sea-level fluctuation.

By analyzing the chemical composition of water samples, cation exchange is discovered. It is characterized that the cation exchange capacity (CEC), expressed by the value of $Na/(Cl+Na)$,[8] increases along with TDS decreasing, and the CEC of IV- and V-aquifer (average 0.65) is greater than that of I–III aquifers (average 0.45).

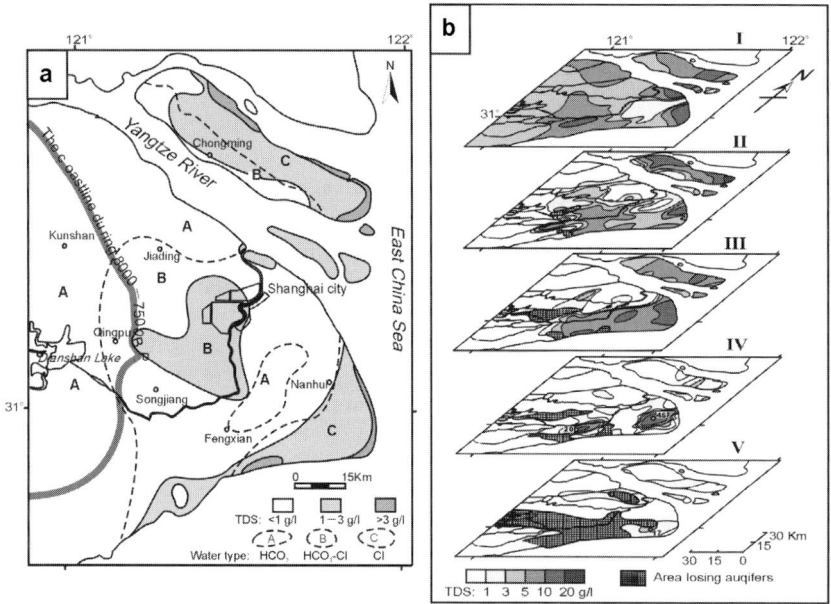

Fig. 1. TDS distribution of groundwater in phreatic aquifer (a) and confined aquifers (b), the coastline during early Holocene is after Yan and Xu.[5]

In addition, we presume that the saltwater in IV- and V-aquifer could formed by means of mixing with upper saltwater through leakage flow, the bases are: (a) marine transgressions from early to middle Pleistocene are very weak and (b) ions composition of samples in the two aquifers shows the characteristics of "resonance" and intergradation (see Fig. 2).

4. Hydrodynamic Evolution of Groundwater

Naturally, the groundwater flow is very slow in study area due to the low hydraulic gradient (ranging from 1/10,000 to 1/100,000) and the finer sediments. In phreatic aquifer, groundwater ridge lies along the line of Jiading–urban district of Shanghai–Nanhui–Fengxia (Fig. 1(a)), which is the highland area. The groundwater renews quickly through dilution by rainwater, and furthermore, also exchanges with surface water in local according to hydrogen isotope.[4] In confined aquifers, the regional groundwater flows from north-western to south-eastern relating with glacis of basal rock. The natural discharge area is located on continental shelf of East

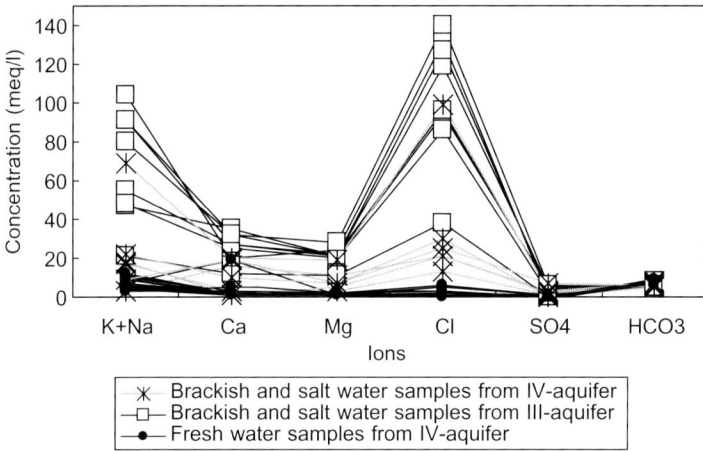

Fig. 2. Ions composition of groundwater samples from III- and IV-aquifer in Minhang and Fengxian area.

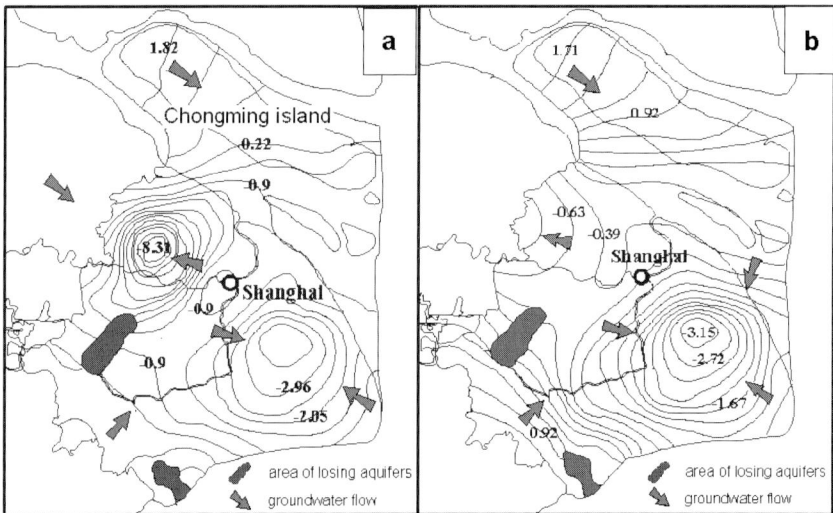

Fig. 3. Simulating result of groundwater flow for II aquifer in 1987 (a) and in 1990 (b).[9]

China Sea. Groundwater exchanges with upper water by leakage flow in local where aquitards is losing.[4]

In order to demonstrate hydrodynamic change, the MODFLOW was applied. The simulated result (see Fig. 3) shows that the water flow field in

Chongming island area represents the regional water flow because of rarely exploitation, and appears remarkably different in the south of estuary. In fact, Shanghai city was the center of heavy extraction in the study area during 1949–1966. After limitation of pumpage and implementation of artificial recharge in order to control land subsidence, groundwater level in the city was rising gradually. Contrarily, more and more groundwater was pumped in adjacent region. These induced the groundwater flow changed from convergent movement to divergent, centered on the urban district of Shanghai. It is obvious that the variety of artificial discharge/recharge intensity is the main driving force for groundwater flow change. Even the flow field change can result in the enlargement/shrinkage of freshwater range in the urban.

5. Groundwater System

Summing up above contents, three groundwater systems in the study could be divided (see Table 2): (a) shallow system. It has never been used due to bad water quality, so groundwater evolves naturally though dilution by rainwater by and large and by surface water in local; (b) middle system. Chemical evolution ways include cation exchange and mixing with the water in adjacent layers by leakage flow in local. This system was already subjected to groundwater overdraft, and the flow field was disturbed to a great extent. Groundwater evolves was droved by human activities; (c) deep system. This system is similar to middle system about the ways of hydrochemical evolution but with more closed hydrodynamic condition and the groundwater ages are older than 16 ka B.P.[10]

Table 2.   Division of groundwater system in the study area.

Groundwater system	Including aquifers	Hydrodynamic	Chemical evolution	Human activities
Shallow system	Phreatic aquifer	Quick renewal of water	Dilution by rainwater	Irrigation
Middle system	I–III	Renewal in local by leakage flow	cation exchange and mixing*	Artificial discharge/ recharge
Deep system	IV–V	Just about logjam	cation exchange	Artificial discharge/ recharge

*Mixing with the water in adjacent layer. The water flow direction is decided by the difference of water table between the adjacent aquifers.

References

1. Z. Y. Chen and D. J. Stanley, *J. Coastal Res.* **11**(3) (1995) 925.
2. Jiangsu Bureau Geol. and Min., Zhejiang Bureau Geol. and Min. and Econ. Geol. Center Shanghai, (eds.), *The Evaluating Report on Hydrogeology and Engineering Geology in Yangtze Delta Region* (not published, 1988), pp. 158–173 (in Chinese).
3. M. G. McDonald and A. W. Hargaugh, *US Geol. Surv. Tech. Water Resour. Invest.* **B6** (1988) 80–100.
4. Z. D. Yuan, *Shanghai Geol.* **9**(2) (1983) 22 (in Chinese, with English abstract).
5. Q. S. Yan and S. Y. Xu (eds.), *Recent Yangtze Delta Deposits* (East China Normal University Press, Shanghai, 1987), p. 438 (in Chinese).
6. Q. B. Min and P. X. Wang, *J. Tongji Univ.* (2) (1979) 109 (in Chinese, with English abstract).
7. Z. Y. Chen, Z. L. Chen and W. G. Zhang, *Quaternary Res.* **47**(2) (1997) 181.
8. A. W. Hunslow (ed.), *Water Quality Data: Analysis and Interpretation* (Lewis Publisher, Boca Raton, 1995), p. 38.
9. Z. Y. Guo, Z. Y. Chen, Z. H. Wang, B. P. Song and Y. Lu, *J. Palaeogeography* **3**(3) (2001) 89.
10. A. B. Hao, K. J. Wang and C. Y. Ha, *Geol. Rev.* **44**(2) (1998) 219 (in Chinese, with English abstract).

FUNCTIONAL CONNECTED AREAS: SEA AND LAND USES INTERACTION

FRANCI STEINMAN*, PRIMOZ BANOVEC and LEON GOSAR
University of Ljubljana, Faculty of Civil and Geodetic Engineering
Jamova 2, SI-1000 Ljubljana, Slovenia
franci.steinman@fgg.uni-lj.si

To a great extent, the land use planning indirectly dictated and still dictates the use of the sea in the narrow coastal belt (since these are the so-called functionally connected lands). The discussion about the sea related issues will have to focus also on the consequences brought by the authorized uses and the applicable legislation on the individual (spatially defined) areas. Regarding the recognized notion of the use of the sea and the water rights, it is first necessary to record and present the actual state of the use of the sea, the rights and the obligations associated to the obtained water rights arising from the national regulations and regulations of the local communities, which relate to an individual area of the authorized use or to the performance of activities.

The modeling approach used for the entire Slovenian coastline is presented. In the article methodology will be presented by which a turn from qualitative assessments (e.g. how the sea is problematic) to the suitable conceptual and quantitative solutions could be performed. The proposed system of marine area use cadastre integrates uses determined by different segments of legislation. With an application of modeling tools this works as a modulus in the development of river basin management plan that includes also coastal zone.

1. Introduction

Historically speaking, coastal areas have always been the development focus of human society and will even in the future remain one of the areas with comparative development advantages in maritime countries. In the Republic of Slovenia, the coastal area is (was) subject to strong development pressures, manifested in constant population growth, urbanization and intensive development of sea-related activities. There are however numerous physical — water resource management and environmental conflicts, but also economic — governance conflicts in use of the terrestrial and marine environment and space.[6] Non-harmonized approaches in sector development plans of coastal urbanism, agriculture, industry, transport, tourism, protection of cultural, and natural heritage, etc. are all reflected in overlapping conditions for use of the sea, coast, and adjacent land. The concentration of functions in the narrow coastal strip, a limited space, is already causing irrational and mutually excluding (economic) uses or functions.

Marine environment and space management was in the past somewhat overshadowed by other contents of water management, mainly focusing on internal (continental) waters, including tributaries of the coastal sea. In fact, even globally, only recently more attention has been given to issues concerning sea management, comprehensively dealing with functional relationship among the sea and continental environments. Although the recently adopted Council Directive establishing a framework for Community action in the field of water policy already covers many issues for determining the action framework for the EU concerning water policy (in short Water Framework Directive), goals concerning sea management within the EU are being listed and regulations for integral management of coastal areas are being prepared.[2] The coastal sea and areas have specific characteristics that cannot be compared with continental water bodies, implying specific approach in pertaining legislature, water resource management and development planning.

To tackle issues tied to the sea it is necessary to first analyze the consequences of legitimized uses and valid legislature on particular (physically defined) areas of the sea.[3] For this reason, these uses affecting the sea were first listed and then physically positioned, since they can practically limit any further uses. The key issue is, that marine area use could be permitted and independently (sectorial) managed by different institutions from different levels of public administration.[5] And here are also the consequences — there is no common list or integration point of national and local legislature. Furthermore, there is no synthesis of consequences of enforcement of other acts applying to particular areas with legitimized use or conducting of activities. This is why we carried out an analysis of physical dimensions of legal regimes, as enforced in valid legislature as well as general and particular government documents.

2. The Legal Regime Term

The legal regime could be defined as a set of legal norms enforced by an act, whereby the method of exercising given rights of use and subsequent obligations (or limitations) is codified for a uniformly defined area. The legal regime is in general applied to property empowerment (e.g. persons of public law) and therefore it can only be enforced when legal basis can be ascertained in a law, which also specifies criteria according to which a legal regime and the area to which it applies can be undoubtedly defined. Simultaneously, the law has to define methods for granting adequate reimbursement.

Legal regimes can be applied upon anything and they address everybody; even a coincidental user (e.g. tourist) and therefore they should become common knowledge. Completely different question for example are, how many inhabitants of the coastal municipalities know, where on the Slovenian coast is subaquatic diving prohibited or even how many visitors to the coast have knowledge about the enforced limitations that have been in effect for ages. Details about legal regimes on properties (buildings, land) should in the future be written as burdens on the property since they affect the scope of rights, obligations, and limitations, which should be known not only to the owner, renter or property manager, but everybody (even potential buyer).

Emphasis should be given to the fact that legal regimes on property, albeit maritime or coastal, in general present limitations of use. They can nevertheless also affect other (e.g. neighboring) maritime and coastal uses, but also other activities, even because of functional ties of the sea with other waters, the sea being a recipient of pollution etc. Despite the stated, open access to the sea has to be granted unprohibitively with general use of the public maritime good (seawater, water area, and coast), which is undoubtedly in public interest.

2.1. Water right and legal acts as an integrative tool — reality check problem

The issue of integration and integrative management is looming the environmental science already for some time. At the same time issue of decision support tools is going alongside with the development of the informatics. The question arise: how could we define a good decision support tool in the light of integrative management? And one of the answers is, that it is able to function in the real world (it survives the reality check).[1] The hardest reality check which could be recognized in the real world is probably the litigation procedure. In the case of conflicts (conflict of opinion, conflict of interest, etc.) the decision made by the mutually recognized arbiter stands for the generally recognized reality check. By this we have clearly shown the importance of the close connection to legal acts in any of the applications considered.

General legal acts as well as individual (or specific) legal acts as individual water rights generally arc, are representing for the legal reality, which is representative also for certain area (i.e. marine area). Reverse connectivity of this reality to the actual physical (also chemical, biological, and other) environment is another link that has to be maintained by the means of established mechanisms like official supervision (inspectors), monitoring and other recognized services which recognitions present a reliable link between the legal reality and actual developments in the real world.

Water rights (individual and general) are therefore essential element for any model that serves as an decision support tool in the field of water management, which is also the fact recognized by many researchers working on the matter. The integration of this dimension in the complex models of water management is still an effort to be made. It is above the scope of the presented article to decompose integration procedures by the means of integration of different aspects (economic aspects, technical aspects, and legal aspects), with essential developments in this field made by.

2.2. *Legal regimes on the sea in the Republic of Slovenia*

Alongside laws on waters and water management, even sectorial laws applied to water can enforce additional areas with legal regimes. The water legislature enforces the well-known water resource protection areas adjacent to water bodies, riparian areas (rivers and streams) and coastal areas (standing water), maintenance strips along water infrastructure (dykes, etc.), flood retention areas, areas for reservoirs, etc. The Water Framework Directive in requesting that protected areas, whose existence is conditioned by the presence of water or water management systems, are listed in water management guidelines, while areas, which should be protected in the future, have to be harmonized with water management plans.

Legal regimes that stem from uses of the sea or the continent with adjacent uses functionally tied to the sea (Fig. 1) are conditioned by particular sectors and their regulations[4]:

- protected area — nature preservation (law on nature, LN),
- protected area — preservation of cultural heritage (law on protection of cultural heritage, LPCH),
- aquatory of ports, anchorages, shipping routes — reservation for shipping (Maritime Law, ML),
- beaches and bathing areas (managed or declared) — prevention from drowning (law on prevention of drowning, LPD),
- fishing reserves and mariculture aquatories — fisheries (law on maritime fishing, LMF).

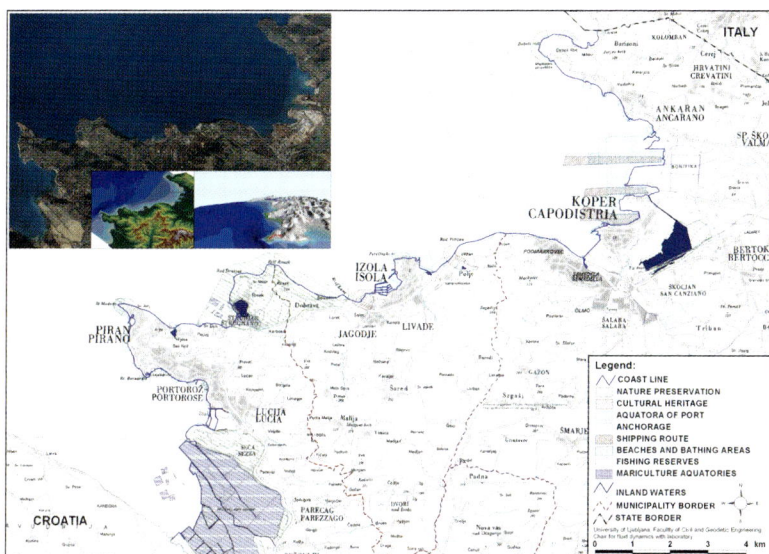

Fig. 1. Areas with enforced legal regimes — tied to use of the sea or as pertaining functional areas of permitted uses and continental activities (map TK50 — Surveying office of the Republic of Slovenia; data on regimes — KMTe, FGG).

Already a very brief summary of prohibitions, limitations or obligations that apply to beaches, bathing areas and aquatories of mariculture enables presentation of mutual interaction of their effects, conflicting uses or even general use of the sea. Abroad the later has the highest priority. Therefore, general use is limited only under extreme conditions (e.g. national defense interests).

For contents shown in Fig. 1, we have to emphasize again that enforced legal regimes determine conduct, even of coincidental users (local population, tourists, etc.). From the remaining parts of the coastline, which are (still) outside areas with applied legal regimes (whereby free access to the sea is possible), one can see that such areas on the Slovenian coast are very few.

3. Conclusions

The presented approach combining in close interaction existing (or foreseen) legal regimes show that the sea surface is not an empty space, therefore enforcement of integral planning of water, land and natural resources (the sea) is essential. This does not mean that expanding planning of existing land use on the continent to the sea would be enough. Abroad corporations are emerging that connect subjects of public and ordinary law, thus creating formal and informal ties between states, local-government, associations of water users, the public, affected individuals, etc. according to principles of partnerships.

A spin-off result of the analysis where the established consequences of poorly thought out enforced legal regimes, which demand serious recollection about the future granting of special rights or limiting areas with legal regimes. The article provides information on the development of the decision support tool that has in its core spatial representation of the legal acts. At the first stage already identification of direct conflicts (actual, potential) is a clear result of the effort. At the same time the created database acts as an unofficial reference point for any considerations about the subject. But the identification of conflicts is not the end stage of the developments as we move towards the more advanced approaches of the optimization of the sustainable marine environment use.

References

1. P. Banovec, L. Gosar and F. Steinman, Coastal zone management plan in slovenia — Marine area uses, User Proceedings from 14, ESRI European User Conference, Munich, Germany, 1999.
2. L. Gosar, Sea area uses in water management, Master of Science Thesis, University of Ljubljana, Faculty of Civil and Geodetic Engineering, Ljubljana, 2000.
3. F. Steinman and L. Gosar, *Integral Management with Maritime and Terrestrial Waters*, (Integralno gospodarenje vodama mora i kopna, Hrvatske vode, Zagreb, Croatia, 2001).
4. F. Steinman, L. Gosar and P. Banovec, *Maritime Legal Systems* (Pravni režimi na moru, Hrvatske vode, Zagreb, Croatia, 2002).
5. P. Burbridge, The guiding principles for a European ICZM strategy, Towards a European strategy for Integrated Coastal Zone Management (ICZM) University of Newcastle, UK, 1999.
6. M. Capobianco, *EU Demonstration Programme on Integrated Coastal Zone Management, Role and Use of Technologies in Relation to ICZM* (Technomare S.p.A., Venezia, Italy, 1999).

IMPROVED ROBUSTNESS AND EFFICIENCY OF THE SCE-UA MODEL-CALIBRATING ALGORITHM

NITIN MUTTIL*,§ and SHIE-YUI LIONG†,¶

*Department of Civil and Structural Engineering
Hong Kong Polytechnic University, Hong Kong
§cenitinm@polyu.edu.hk

†Tropical Marine Science Institute
National University of Singapore, Singapore
¶tmslsy@nus.edu.sg

The Shuffled Complex Evolution (SCE-UA) has been used extensively and proved to be a robust and efficient global optimization method for the calibration of conceptual models. In this study, we propose two enhancements to the SCE-UA algorithm to improve its exploration and exploitation of the search space. A strategically located initial population is used to improve the exploration capability and a modification to the downhill simplex search method enhances its exploitation capability. This enhanced version of SCE-UA is tested on a suite of test functions and it is observed that the strategically located initial population drastically reduces the number of failures and the modified simplex search also leads to significant reduction in the number of function evaluations to reach the global optimum, when compared to the original SCE-UA. Thus, the two enhancements further improve the robustness and efficiency of the SCE-UA algorithm.

1. Introduction

With the advent of digital computers, a generation of models known as conceptual models has been developed. The successful application of a model heavily depends on how well it is calibrated. There is a substantial body of research documenting problems encountered during model calibration, especially with conceptual models.[1-3] Duan[1] pointed out five characteristics that complicate the optimization of conceptual models. The most important of these characteristics is the presence of multiple optima.

To deal with the problem of multiple local minima, global search methods are applied. These methods are global in the sense that they constitute a parallel search of the search space (as opposed to a point by point search) by using a population of potential solutions. This capability of such techniques for effective "exploration" of the search space makes them less probable to get trapped into local minima. Popular global search methods are the so-called population-evolution-based search strategies such as the Shuffled Complex Evolution[1] (SCE-UA) and the Genetic Algorithm.[4] This study deals with significant improvement of the robustness and efficiency of the SCE-UA. Thus, a brief introduction to the SCE-UA algorithm is first presented in the next section. This is followed by the two proposed enhancements and then the comparison of the enhanced SCE-UA with its original counterpart on popular test functions is presented.

2. The Shuffled Complex Evolution

The SCE-UA combines the best features of "multiple complex shuffling" and "competitive evolution" based on the downhill simplex search method.[5] The use of multiple complexes and their periodic shuffling provide a more effective exploration of different promising regions of attraction within the search space. This effective exploration is coupled with evolution of each simplex, which is provided by the simplex search method. This "competitive evolution" of the simplexes provides effective exploitation within the search space. Thus, the SCE-UA achieves a superior balance between exploration and exploitation as compared to other population-evolution-based search strategies. For a lucid explanation on the details of the algorithm, the reader is referred to Ref. 6.

A number of studies have been conducted to compare the SCE-UA and other global and local search procedures for model calibration.[1-3,7] These studies have demonstrated that the SCE-UA method is an effective and efficient search algorithm.

3. The Proposed Enhancements

Various population-evolution-based search strategies, including the SCE-UA use a random data generator to generate the initial population. As the search proceeds, the population converges toward an optimum in one of the many possible regions of attraction. If this region of attraction does not contain the global optimum, then the search converges to a local optimum. The reason for such local minimum convergence could be insufficiently large initial population size or an initial population that is not well spread in the search space.

Thus, with the aim of having an initial population of points that are well spread in the search space, a scheme to strategically locate the initial

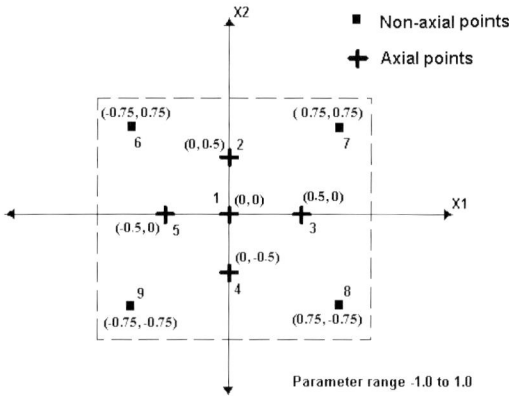

Fig. 1.   Strategically located initial population.

population of points was proposed in a previous study.[8] Figure 1 shows
the locations of points of initial population, suggested by the proposed
scheme, for a 2-dimensional (2D) search space. The strategically located
initial population was shown to increase the exploratory capability of the
search algorithm.

In this study, with the aim of improving the exploitative capability,
a modification to the simplex search method is presented. In the simplex
search method employed in the original SCE-UA, the new points are generated by reflecting (or contracting) the worst point (X_w in Fig. 2) in a
simplex about the centroid of the remaining points (X_c). We propose to
shift the newly generated reflected (or contracted) point towards the best

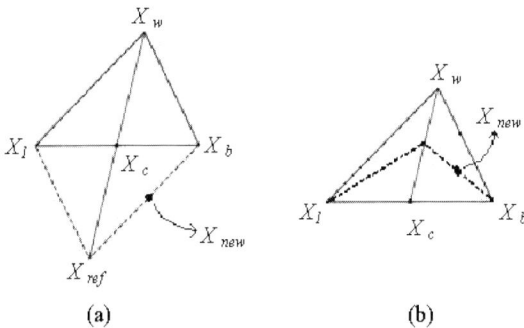

Fig. 2.   The reflection and contraction steps when $\theta = 0.5$.

point in the simplex (X_b), with the aim of directing the simplex toward the optimum using a lesser number of function evaluations. Thus, not only the worst point, but also the best point in a simplex is used, making better use of the already available information.

The reflected and contracted points are shifted toward the best point using a parameter theta, θ, which is defined as below for reflection and contraction, respectively.

$$X_{\text{new}} = ((1.0 - \theta) * X_{\text{ref}}) + (\theta * X_b), \tag{1}$$

$$X_{\text{new}} = ((1.0 - \theta) * X_{\text{con}}) + (\theta * X_b), \tag{2}$$

where X_{ref} is the reflected point, X_{con} the contracted point and X_b is the best point in the simplex. The parameter θ can take values between 0.0 and 1.0. The higher its value, the more is the exploitation pressure, since the new point (X_{new}) moves closer to the best point (X_b). For $\theta = 0.5$, the new point, X_{new} is in the middle of the reflected (or contracted) point and the best point (X_b), which is shown in Figs. 2(a) and (b), for reflection and contraction steps, respectively.

4. Experiment on Test Functions

In this section, a performance comparison between the original SCE-UA and the enhanced SCE-UA on a series of test functions is presented.

For the two types of SCE-UA algorithms, performance criteria used are: (i) the number of failures (NF) out of 100 trials; and (ii) the average number of function evaluations (AFE) resulting from successful trials. NF measures robustness while AFE describes the efficiency of the algorithm.

The stopping criteria used are as follows. A trial is deemed a success as soon as the best function value in the sample became less than 10^{-3}. However, if the trial reached 25,000 function evaluations without reducing the best function value below 10^{-3}, the trial was deemed a failure. Exception to the stopping criterion of 25,000 function evaluations are the Neumaier No. 3 function and the Griewank 10D function, which being 10-dimensional functions, are expected to require higher number of function evaluations. As such, the maximum function evaluation for these two test functions is set to 50,000.

The results of the comparison are presented in Table 1. For the enhanced version of the algorithm, the value of θ that gave the best results is also presented in Table 1. It is seen that best results are obtained when θ is

Table 1. Comparison of the original and enhanced SCE-UA algorithms.

Function name	No. of complexes; population size	Original SCE-UA		Enhanced SCE-UA		
		NF	AFE	θ	NF	AFE
Goldstein-Price (2D)	(2; 10)	2	162	0.5	0	86
Rosenbrock (2D)	(2; 10)	0	274	0.2	0	214
6-hump camelback (2D)	(2; 10)	0	162	0.5	0	87
Rastrigin (2D)	(2; 10)	34	340	0.0	20	303
Griewank (2D)	(2; 10)	12	355	0.2	9	289
Schwefel (2D)	(2; 10)	53	257	0.3	14	177
Shekel (4D)	(3; 27)	23	494	0.2	0	415
Hartman (6D)	(6; 78)	10	673	0.4	0	469
Neumaier (10D)	(50; 1,050)	0	20,989	0.6	0	9,760
Griewank (10D)	(50; 1,050)	0	28,843	0.6	0	15,438

Note: NF: number of failures in 100 runs and AFE: average function evaluations.

in the range 0.1–0.5 and values higher than this lead to an increase in the number of failures. It is clearly seen that the proposed enhancements significantly reduce the NF and also the AFE of the SCE-UA algorithm. Thus, the proposed enhancements lead to significant improvement in the robustness and efficiency of the SCE-UA.

5. Conclusions

The present study proposes two enhancements to the SCE-UA model-calibrating algorithm, which is compared with the original SCE-UA on a suite of test functions. A scheme to systematically, instead of randomly, generating the initial population leads to much better exploration and a significant reduction in the number of failures. The second enhancement, a modification to the downhill simplex search method leads to enhanced exploitation, which in turn leads to a significant reduction in the function evaluations to reach the global optimum. Thus, the two proposed enhancements improve the robustness and efficiency of the SCE-UA model-calibrating algorithm.

Acknowledgments

The authors wish to thank Dr. Q. Duan of the NOAA (National Oceanic and Atmospheric Administration, USA) for kindly providing the source code for the SCE-UA (version 2.2).

References

1. Q. Duan, *Wat. Res. Res.* **28** (1992).
2. T. Y. Gan and G. F. Biftu, *Wat. Res. Res.* **32** (1996).
3. G. Kuczera, *Wat. Res. Res.* **33** (1997).
4. Q. J. Wang, *Wat. Res. Res.* **27** (1991).
5. J. A. Nelder and R. Mead, *J. Comput.* **7** (1965).
6. Q. Duan, S. Sorooshian and V.K. Gupta, *J. Hydrol.* **158** (1994).
7. M. Franchini, G. Galeati and S. Berra, *J. Hydrol. Sci.* **43** (1998).
8. N. Muttil and S. Y. Liong, *J. Hydraul. Eng.* **130** (2004).